BIOINFORMATICS: AN INTRODUCTION

Computational Biology

VOLUME 3

BIOINFORMATICS:
AN INTRODUCTION

by

JEREMY J. RAMSDEN

Cranfield University,
Cranfield, United Kingdom

KLUWER ACADEMIC PUBLISHERS
DORDRECHT / BOSTON / LONDON

A C.I.P. Catalogue record for this book is available from the Library of Congress.

ISBN 1-4020-2141-0 (HB)
ISBN 1-4020-2143-7 (E-book)

Published by Kluwer Academic Publishers,
P.O. Box 17, 3300 AA Dordrecht, The Netherlands.

Sold and distributed in North, Central and South America
by Kluwer Academic Publishers,
101 Philip Drive, Norwell, MA 02061, U.S.A.

In all other countries, sold and distributed
by Kluwer Academic Publishers,
P.O. Box 322, 3300 AH Dordrecht, The Netherlands.

Printed on acid-free paper

Printed in the Netherlands.

Mi a tudvágyat szakhoz nem kötők,
Átpillantását vágyuk az egésznek.

IMRE MADÁCH

Contents

PREFACE

This little book attempts to give a self-contained account of bioinformatics, so that the newcomer to the field may, whatever his point of departure, gain a rather complete overview. At the same time it makes no claim to be comprehensive: the field is already too vast—and let it be remembered that although its recognition as a distinct discipline (i.e. one after which departments and university chairs are named) is recent, its roots go back a long time.

Given that many of the newcomers arrive from either biology or informatics, it was an obvious consideration that for the book to achieve its aim of completeness, large portions would have to deal with matter already known to those with backgrounds in either of those two fields, i.e. in the particular chapters dealing with them the book would provide no information for them. Since such chapters could hardly be omitted, I have tried to consider such matter in the light of bioinformatics as a whole, so that even the student ostensibly familiar with it could benefit from a fresh viewpoint.

In one regard especially this book cannot be comprehensive. The field is developing extraordinarily rapidly and it would have been artificial and arbitrary to take a snapshot of the details of contemporary research. Hence I have tried to focus on a thorough grounding of concepts, which will enable the student not only to understand contemporary work, but should also serve as a springboard for his or her own discoveries. Much of the raw material of bioinformatics is open and accessible to all via the internet, powerful computing facilities are ubiquitous, and we may be confident that vast tracts of the field lie yet uncultivated. This accessibility extends to the literature: research papers on any topic can usually be found rapidly by an internet search, and therefore I have not aimed at providing a comprehensive bibliography.

In bioinformatics, so much is to be done, the raw material to hand is already so vast and vastly increasing, and the problems to be solved are so important (perhaps the most important of any science at present) we may be entering an era comparable to the great flowering of quantum mechanics in the first three decades of the twentieth century, during which there were periods when practically every doctoral thesis was a major breakthrough. If this book is able to inspire the student to take up some of the challenges, then it will have accomplished a large part of what it sets out to do.

Indeed, I would go further to remark that I believe that there are still comparatively simple things to be discovered, and that many of the present directions of work in the field may turn out not to be right. Hence, at this stage in its development the most important thing is to facilitate that viewpoint that will facilitate new discoveries. This belief also underlies the somewhat more detailed coverage of the biological processes in which information processing in nature is embodied than might be considered customary.

A work of this nature depends on a long history of interactions, discussions and correspondence with many present and erstwhile friends and colleagues, some of whom, sadly, are no longer alive. I have tried to reflect some of this debt in the citations. Furthermore, many scientific subjects and methods other than those mentioned in the text had to be explored before the ones best suited to the purpose of this work could be selected, and my thanks are due to all those who helped in these preliminary studies. I should like to add an especial word of thanks to Victoria Kechekhmadze for having so ably drawn the figures.

CRANFIELD
January 2004

Chapter 1

Introduction

INFORMATION IS CENTRAL to life. The principle enunciated by Crick, that information flows from the gene (DNA) to the protein, occupies such a key place in modern molecular biology that it is frequently referred to as the "central dogma": DNA acts as a template to replicate itself, DNA is transcribed into RNA, and RNA is translated into protein.

The mission of biology is to answer the question "What is life?" For many centuries, the study of the living world proceeded by examination of its external characteristics, i.e. of phenotype (including behaviour). This led to Linnaeus' hierarchical classification. A key advance was made about one hundred and fifty years ago when Mendel established the notion of an unseen heritable principle. Improvements in experimental techniques lead to a steady acceleration in the gathering of facts about the components of living matter, culminating in Watson and Crick's discovery of the DNA double helix half a century ago, which ushered in the modern era of molecular biology.

The mission of biology remained unchanged during these developments, but knowledge about life became steadily more detailed. As Sommerhoff remarked, "To put it naïvely, the fundamental problem of theoretical biology is to discover how the behaviour of myriads of blind, stupid, and by inclination chaotic, atoms can obey the laws of physics and chemistry, and at the same time become integrated into organic wholes and into activities of such purpose-like character." Since he wrote those words, experimental molecular biology has advanced far and fast, yet the most vital question of all, "what is life?", remains a riddle.

It is a curious fact that although "information" figured so prominently in the central dogma, which is indissociable from the modern era of molecular biology, the concept of information continued to receive extremely cursory treatment in molecular biology textbooks. Even today, it rarely gets a mention, and the word may not even appear in the index. On the other hand

1

whole chapters are devoted to energy and energetics, which, like information, is another fundamental, irreducible concept. Although the doctoral thesis of Shannon, one of the fathers of information theory, was entitled "An algebra for theoretical genetics", apart from genetics, biology remained largely untouched by developments in information science.

One might speculate on why information was placed so firmly at the core of molecular biology by one of its pioneers. During the preceding decade, there had been tremendous advances in the theory of communication—the science of the transmission of information. Shannon published his seminal paper on the mathematical theory of communication only a few years before Watson and Crick's work. The notion of the sequence of DNA bases as message with meaning seemed only natural, and the next major development, the establishment of the genetic code with which the DNA sequence could be transformed into a protein sequence, was cast very much in the language and concepts of communication theory. More puzzling is that there was not a more vigorous interchange between the two disciplines. Probably the lack of extensive datasets, and the lack of powerful computers, which would have made the necessary calculations intolerably tedious, or simply too long, provides sufficient explanation for this neglect—and hence, now that both these requirements (datasets and powerful computers) are being met, it is not surprising that there is a great revival in the application of information ideas to biology. One may indeed hope that this revival will at last lead to a real answer being advanced in response to the vital question "what is life?": in other words, information science is perhaps the 'missing discipline' that, along with the physics and chemistry already being brought to bear, is needed to answer the question.

What is bioinformatics?

The term 'bioinformatics' seems to have been first used in the mid 1980s in order to describe the application of information science and technology in the life sciences. The definition was at that time very general, covering everything from robotics to artificial intelligence. Later, bioinformatics came to be somewhat prosaically defined as "the use of computers to retrieve, process, analyse and simulate biological information". An even narrower definition was "the application of information technology to the management of biological data". Such definitions fail to capture the centrality of information in biology. If indeed information the most fundamental concept underlying biology, and bioinformatics is the exploration of all the ramifications and implications of that fundament, then bioinformatics is excellently positioned to revive consideration of the vital question "what is life?" A more appropriate definition of bioinformatics is therefore "the science of how information is generated, transmitted, received and interpreted in biological systems", or, more succinctly, "the application of information science to biology".

The emergence of information theory by the middle of the twentieth century enabled the creation of a formal framework within which information could be quantified. To be sure, the theory was, and to some extent still is, incomplete, especially regarding those aspects going beyond the merely faithful transmission of messages, and enquiring about, and even quantifying, the meaning and significance of messages.

In parallel to these developments other advances, including the development of the idea of algorithmic complexity, with which the name of Kolmogorov is associated, allowed a number of other crucial clarifications to be made, including the notion that randomness is minimally informative. The DNA sequence of a living organism must depart in some way from randomness, and the study of these departures could be said to constitute the core of bioinformatics.

What can bioinformatics do?

In a very short interval 'bioinformatics' has become an extremely active research field. Although it began with sequence comparison (which is a subbranch of the study of the nonrandomness of DNA sequences) it now encompasses a far wider spread of activity, which truly epitomizes modern scientific research: (a) it is highly interdisciplinary, requiring at least mathematical, biological, physical and chemical knowledge; its implementation may furthermore require knowledge of computer science, chemical engineering, biotechnology, medicine, pharmacology, etc.; and (b) there is little distinction between the work carried out in the public domain, either in academic institutions (universities) or state research laboratories, or privately by commercial firms.

The handling and analysis of DNA sequences remains one of the prime tasks of bioinformatics. This topic is usually divided into two parts, functional genomics, which seeks to determine the rôle of the sequence in the living cell, either as a transcribed and translated unit (i.e. a protein) or as a regulatory motif, etc., and comparative genomics, in which the sequences from different organisms, or even different individuals, are compared in order to determine ancestries and correlations with disease. Clearly the comparison of unknown sequences with known ones can also help to elucidate function; both parts are concerned with the finding of patterns or *regularities*—which is the core of all scientific work. One can feel that it is fortunate (for scientists) that life is in some sense encapsulated in such a highly formalized object as a sequence of symbols.

The requirement of entire genomes to feed this search has led to tremendous advances in the technology for rapid sequencing, which in turn has put new demands on informatics for interpreting the raw output of a sequencer. If a DNA sequence is the message, then functional genomics is already concerned with the meaning of the message, and in turn this has led to the experimental analysis of the RNA transcripts (the transcriptome) and the repertoire of expressed proteins (the proteome), each of which presents fresh informatics challenges. They have themselves spawned interest in the products of protein activity—saccharides (glycomics), lipids (lipidomics) and metabolites (metabolomics). All these '– omics' are considered to be part of bioinformatics and are covered in this book; some closely related topics, such as chemical genomics (defined as the large use of small molecules to study the function of gene products), and computational biology (defined as the application of quantitative and analytical techniques to model biological systems), will not be covered. This omission includes the impressive attempts of Holland, Ray and others to model essential features of life—speciation and evolution—entirely *in silico* using digital organisms, i.e. computer programs able to self-replicate, mutate etc.

It would be perfectly reasonable to include neurophysiology in bioinformatics, since it deals with how information is generated, transmitted, received and interpreted in the brain, i.e. it corresponds precisely with our definition above, but although in the future it may well come to be considered as part of bioinformatics, at present it is a vast field in its own right, with its own independent traditions, and we shall not consider it here. The same remarks apply to the whole science of human communication.

The book is organized into three main parts. Part I deals, essentially heuristically, with the concept of information, and some essential basic knowledge including elements of combinatorics and probability theory. Part II is a compact primer on biology, both molecular and organismal. It includes formal aspects of mechanism, whether living or not, such as regulation and adaptation. Part III deals with applications, i.e. the areas of active current research, including genomics, proteomics and interactomics (the study of the repertoire of molecular interactions in a cell). Topics such as practical programming, or database handling, and left out since there are already several excellent books available covering them.

Although the gene has been at the heart of bioinformatics from the beginning, the main challenge seems now to lie in understanding the functional relationships between biological objects beyond those which are encoded in the nucleotide sequence. This zone is called epigenetics, and we are only just entering it. It still appears mostly formless, amorphous, and if there are clues to its structure in the nucleotide sequence, they remain as yet largely hidden from us.

Attention should also be called to the fact that for various reasons, including experimental ones, the usual procedure in the physical sciences, which is first to assign numbers to the phenomenon under investigation, and then to manipulate the numbers according to the usual rules of mathematics, both operations being publicly declared and publicly accessible, is often confounded in the biological sciences. Bioinformatics may be able to provide the needed quantification over the vast tracts of biology where it is so sorely needed.

One consequence of the apparent reluctance of experimenters in the biological sciences to assign numbers to the phenomena they investigate is that the experimental literature is very wordy and hence voluminous, so much so that a sub-branch of bioinformatics called text mining has grown up, whose aim is to automatically extract information from published articles, from which, for example, the association of a pair of genes can be inferred. The techniques involved are essentially the same as those involved in searching for genes in a DNA sequence, and hence will not be considered separately in this book.

Activity in a new field begins firstly with the advanced researcher, later it becomes material suitable for doctoral theses, and finally becomes part of undergraduate studies. Bioinformatics seems to be on the threshold of the shift into undergraduate work. The enormous virgin fields opened up by the sequencing of the entire DNA of organisms has imparted tremendous impetus and urgency, and practitioners are now required at every level, from the implementation of the latest findings in medicine and ecology, to the continued pushing back of the frontiers of knowledge.

Part I

Information

Chapter 2

The nature of information

What is information? We have already asserted that it is a profound, primitive (i.e. irreducible) concept. Dictionary definitions include "(desired) items of knowledge." For example, one wishes to know the length of a piece of wood. It appears to be less than a foot long, so we measure it with our desktop ruler marked off in inches, with the result, let us say, "between six and seven inches." This result is clearly an item of desired knowledge, hence information. We shall return to this example later. Another definition is "fact(s) learned about something," implying that there is a definable object to which the facts are related, suggesting the need for context and meaning. A further definition is "what is conveyed or represented by a particular arrangement of things"; the dots on the head of a matrix printer shape a letter, the bar code on an item of merchandise represents facts about the nature, origin and price of the merchandise, and a sequence of letters can convey a possibly infinite range of meanings. A thesaurus gives as synonyms "advice, data, instruction, message, news, report." Finally we have "a mathematical quantity expressing the probability of occurrence of a specific sequence of symbols or impulses etc. as aganst that of other sequences (i.e. messages)." This definition links the quantification of information to a probability, which, as we shall see, plays a major rôle in the development of the subject.

We also note that 'information science' is defined as the "study of processes for storing and retrieving information", and 'information theory' as the "quantitative study of transmission processes for storing and retrieving of information by signals," i.e. it deals with the mathematical problems arising in connexion with the storage, transformation and transmission of information. This forms the material for Chapter 3. Etymologically, the word 'information' comes from the Latin *forma*, form, from *formare*, give shape to, describe.

Most information can be reduced to the response, or series of responses, to a question, or series of questions, admitting only yes or no as an answer. We call these yes/no, or dichotomous, questions. Typically, interpretation depends heavily on context. Consider a would-be passenger racing up to a railway station. His question 'has the train gone?' may indeed be answered by 'yes' or 'no'—although, in practice, a third alternative, 'don't know', may be encountered. At a small wayside station, with the traveller arriving within five minutes of the expected departure time of the only train scheduled within the next hour, the answer (yes or no) would be unambiguous and will convey exactly one bit of information, as will be explained below. If we insist on the qualification 'desired', an unsolicited remark of the stationmaster, "the train has gone", may or may not convey information to the hopeful passenger. Should the traveller have seen with his own eyes the train depart a minute before, the stationmaster's remark would certainly not convey any information.

Consider now a junction at which, after leaving the station, the lines diverge in three different directions. The remark "the train has gone", assuming the information was desired, would still convey one bit of information, but by in addition specifying the direction, viz. "the train has gone to X", or "the train to X has gone", 'X' being one of the three possible destinations, the remark would convey $\log_2 3 = 1.59$ bits of information, this being the average number of questions admitting yes/no answers required to specify the fact of departure to X, as opposed to either of the two other directions.

This little scenario illustrates several crucial points:

1. Variety exists. In a formless, amorphous world there is no information to convey.

2. The amount of information received depends on what the recipient knows already.

3. The amount of information can only be calculated if the set of possible messages (responses) has been predefined.

Dichotomous information often has a hierarchical structure. For example, on a journey, a selection of direction has to be made at every crossroad. Given an ultimate destination, successive choices are only meaningful on the basis of preceding ones. Consider also an infant, who 'chooses' (according to its environment) which language it will speak. As an adolescent, he chooses a profession, again with an influence from the environment, and in making this choice knowledge of a certain language may be primordial. As an adult there will be further career choices, which will usually be intimately related to the previous choice of a profession.

Let us now reexamine the measurement of the length of a stick. It must be specified in advance that it does not exceed a certain value—say one foot. This will suffice to allow an appropriate measuring tool to be selected. If all we had was a measuring stick exactly one foot long, we could simply ascertain whether the unknown piece was longer or shorter, and this information would provide one bit of information, if any length was *a priori* possible for the unknown piece.

Suppose, however, that the measuring stick is marked off in one inch divisions. If the probabilities p of the unknown piece being any particular length l (measured to the nearest inch), with $0 < l \leq 12$, were *a priori* equal, i.e. $p = \frac{1}{12}$ for each possible length, then the information produced by the measurement equals $\log_2 12 = 3.59$ bits, this being the average number of questions admitting yes/no answers required to specify the length to the nearest inch, as the reader may verify. On the other hand, were we to have some prior information, according to which we had good reason to suppose the length to be close to nine inches (perhaps we had previously requested the wood to be chopped to that length), the probabilities of the lengths 8, 9 and 10 inches would perhaps be 0.25 each, and the sum of all the others would be 0.25. The existence of this prior knowledge would somewhat reduce the quantity of information gained from the measurement, namely to $\frac{3}{4} \log_2 4 + \frac{1}{4} \log_2 36 = 2.79$ bits. Should the ruler have been marked off in tenths of an inch, the measurement would have yielded considerably more information, namely $\log_2 120 = 6.91$ bits, assuming all the probabilities of the wood being any particular length to be equal, i.e. $\frac{1}{120}$ each.

Variety

One of the most striking characteristics of the natural, especially the living, world around us is its variety. This variety stands in great contrast to the world studied by the methods of physics and chemistry, in which every electron and every proton (etc.) in the universe are presumed to be identical, and we have no evidence to gainsay this presumption. Similarly, every atom of helium (^{4}He) is similar to every other one, and indeed it is often emphasized that chemistry could only make progress as a quantitative science after the realization that pure substances were necessary for the investigation of reactions etc., such that a sample of naphthalene in a laboratory in Germany would behave in precisely the same way as one in Japan.

If we are shown a tray containing balls of three colours, red (r), blue (b) and white (w), we might reasonably assert that the variety is three. Hence one way to quantify variety is simply to count the number of different kinds of objects. Thus the variety of the sets $\{r, b, w\}$, and $\{r, b, b, r, w, r, w, w, b\}$

is equal to three; the set $\{r, r, w, w, w\}$ has a variety of only two, etc. The objects considered should of course be in the same category; i.e. if the category were specified as 'ball', then we would have difficulty if the tray also included a banana and an ashtray. But one could then redefine the category.

If there were only one kind of ball, say red, then our counting procedure would yield a variety of one. It is more natural, however, to say that there is no variety if all the objects are the same, suggesting that the logarithm of the number of objects is a more reasonable way to quantify variety. If all the objects are the same, the variety is then zero. We are of course at liberty to choose any base for the logarithm; if the base is 2, then conventionally the variety is given in units of bits, a contraction of *binary digit*. Hence two kinds of objects have a variety of $\log_2 2 = 1$ bit, and three kinds give $\log_2 3 = \frac{\log_{10} 3}{\log_{10} 2} = \frac{0.477}{0.301} = 1.58$ bits. The variety in bits is the average number of yes/no questions required to ascertain the number of different kinds of objects, or identify the kind of any object chosen from the set.[1]

The Shannon index

The formula that we used to determine the quantity I of information delivered by a measurement that fixes the result as one out of n equally likely possibilities, each having a probability $p_i, i = 1, \ldots, n$ equal to $1/n$, was

$$I = -\log_2 p = \log_2 n \ . \tag{2.4}$$

It is called Hartley's formula. If the base of the logarithm is 2, then the formula yields numerical values in bits. Where the probabilities of the different alternatives are not equal, then a weighted mean must be taken:

$$I = -\sum_{i=1}^{n} p_i \log p_i \ . \tag{2.5}$$

---

[1]This primitive notion of variety is related to the diversity measured by biometricians concerned with assessing the variety of species in an ecosystem (biocoenosis). Diversity D is essentially variety weighted according to the relative abundances (i.e. probability p_i of occurrence) of the N different types, and this can be done in different ways. Parameters in use by practitioners include:

$$D_0 \ = \ N \quad \text{(no weighting)} \tag{2.1}$$

$$D_1 \ = \ \exp(I) \quad \text{(the exponential of the Shannon index)} \tag{2.2}$$

$$D_2 \ = \ 1/\sum_{i=1}^{N} p_i^2 \quad \text{(the reciprocal of Simpson's index)} \ . \tag{2.3}$$

This generalization is called the Shannon or Shannon-Wiener index. In other words, the quantity of information is weighted logarithmic variety. Note that the information given by equation (2.5) is always less than that given by the equiprobable case (2.4). This follows from Jensen's inequality.[2]

Why is the negative of the sum taken? I in fact represents the *gain* of information due to the measurement. In general,

$$\text{gain (in something)} = \text{final value} - \text{initial value} . \tag{2.7}$$

The initial value represents the uncertainty in the outcome *prior* to the measurement. Shannon takes the *final* value, i.e. the result of the measurement, to be a single value with variety one, hence using (2.5) $I = 0$ after the measurement. That is, he considers the result to be known with certainty once it has been delivered. Hence it is considered to have zero information, and it is in this sense that an information processor is also an information annihilator. Wiener considers the more general case in which the result of the measurement could be less than certain, e.g. still a distribution, but narrower than the one measured.

The gain of information I is equivalent to the removal of uncertainty; hence information could be defined as "that which removes uncertainty". It corresponds to the reduction of variety perceived by an observer, and is inversely proportional to the probability of a particular value being read, or a particular symbol (or set of symbols) being selected, or, more generally, is inversely proportional to the probability of a message being received and remembered.

Example. An $N \times N$ grid of pixels, each of which can be either black or white, can convey at most $-\sum_i^{N^2} \frac{1}{2} \log_2 \frac{1}{2}$ bits of information. This maximum is achieved when the probability of being either black or white is equal.

Equations (2.4) and (2.5) possess the properties that one may reasonable postulate should be possesses by a measure of information, namely

1. $I(E_{NM}) = I(E_N) + I(E_M)$ for $N, M = 1, 2, \ldots$;

---

[2]If $g(x)$ is a convex function on an interval (a, b), if $x_1, x_2, \ldots, x_n$ are arbitrary real numbers $a < x_k < b$ and if $w_1, w_2, \ldots, w_n$ are positive numbers with $\sum_{k=1}^{n} w_k = 1$, then

$$g(\sum_{k=1}^{n} w_k x_k) \le \sum_{k=1}^{n} w_k g(x_k) . \tag{2.6}$$

(2.6) is then applied to the convex function $y = x \log x (x > 0)$ with $x_k = p_k, w_k = 1/n$ $(k = 1, 2, \ldots, n)$ to get $I(p_1, p_2, \ldots, p_n) \le \log n$.

2. $I(E_N) \leq I(E_{N+1})$;

3. $I(E_2) = 1$.

Example.   How much information is contained in a sequence of DNA? If each of the four bases are chosen with equal probability, i.e. $p = \frac{1}{4}$, the information in a decamer is $10 \log_2 4 = 20$ bits. It is the average number of yes/no questions that would be needed to ascertain the sequence. If the sequence were completely unknown before questioning, this is the gain in information. Any constraints imposed on the assembly of the sequence—for example, a rule that 'AA' is never followed by 'T', will lower the information content of the sequence, i.e. the gain in information upon receiving the sequence, assuming that those constraints are known to us. Some proteins are heavily constrained; the antifreeze glycoprotein (alanine-alanine-threonine)$_n$ could be specified by the instruction 'repeat AAT n times', i.e. much more compactly than writing out the amino acid sequence in full, and the information gained upon being informed of the sequence using the compact instruction is correspondingly little.

Thermodynamic entropy

One often encounters the word 'entropy' used synonymously with information (or its removal). Entropy (S) in a physical system represents the ability of a system to absorb energy eithout increasing its temperature. Under isothermal conditions (i.e. constant temperature T),

$$dQ = T dS \qquad (2.8)$$

where dQ is the heat that flows into the system. In thermodynamics, the internal energy E of a system is formally defined by the First Law as the difference between the heat and dW the work done by the system:

$$dE = dQ - dW . \qquad (2.9)$$

The only way that a system can absorb heat is by becoming more disordered. Hence entropy is a measure of disorder. Starting from a microscopic viewpoint, entropy is given by the famous formula inscribed on Boltzmann's tombstone

$$S = k_B \ln W \qquad (2.10)$$

where k_B is Boltzmann's constant and W is the number of (micro)states available to the system. An 'amount of information' of $\log_2 W$ bits is required to specify one particular microstate (assuming that all microstates have the same probability of being occupied) according to Hartley's formula: specification of a particular microstate removes that quantity of uncertainty. Hence thermodynamical entropy (2.8), statistical mechanical entropy (2.10) and the Hartley or Shannon index only differ from each other by numerical constants.

Although the set of positions and momenta of the molecules in a gas at a given instant can thus be considered as information, within a microscopic interval (between atomic collisions, of the order of 0.1 ps) this set is forgotten and another set is realized. The positions and momenta constitute microscopic information; the quantity of macroscopic (remembered) information is zero. In general, the quantity of macroinformation is far less than the quantity of (forgotten) microinformation, but the former is far more valuable.[3]

In the world of engineering, this state of affairs has of course always been recognized. One does not need to know the temperature (within reason!) in order to design a bridge or a mechanism. The essential features of any construction are found in a few large-scale correlated motions; the vast number of uncorrelated, thermal degrees of freedom are generally unimportant.

Symbol and word entropies

The Shannon index (2.5) gives the average information per symbol; an analogous quantity I_n can be defined for the probability of n-mers (n-symbol 'words'), whence the differential entropy $\tilde{I}_n$,

$$\tilde{I}_n = I_{n+1} - I_n , \qquad (2.11)$$

whose asymptotic limit ($n \to \infty$) Shannon calls 'entropy of the source', is a measure of the information in the $(n+1)$th symbol, assuming the n previous ones are known. The decay of $\tilde{I}_n$ quantifies correlations within the symbolic sequence, i.e. memory.

---

[3]'Forgetting' implies decay of information; what does 'remembering' mean? It means to bring a system to a defined stable state, i.e. one of two or more states, and the system can only switch to another state under the influence of an external impulse. The physical realization of such systems implies a minimum of several atoms: as a rule a single atom, or a simple small molecule, can exist in only one stable state. Among the smallest molecules fulfilling this condition are sugars and amino acids, which can exist in left- and right-handed chiralities. Note that many biological macromolecules and supermolecular assemblies can exist in several stable states.

2.1   Structure and quantity

In our discussion so far we have tacitly assumed that we know *a priori* the set from which the actual measurement will come. In an actual physical experiment, this is like knowing from which dial we shall take readings of the position of the pointer, etc., and may comprise all the information required to construct and use the meter, which is far more than that needed to formally specify the blueprints and circuit diagram: it would also have to include blueprints for the machinery needed to make the mechanical and electronic components, and for manufacturing the required materials from available matter, etc. In many cases we do not need to concern ourselves about all this, because we are only interested in the gain in information (i.e. loss of uncertainty) obtained by receiving the result of the dial reading, which is given by equation (2.5). The information pertinent to the construction of the experiment usually remains the same, and hence cancels out (equation 2.7). In other words, the Shannon-Weaver index is strictly concerned with the metrical aspects of information, and not with its structure, i.e. it is a measure of that part of the information that is structured.

2.1.1   The generation of information

Prior to carrying out an experiment, or an observation, there is objective uncertainty due to the fact that several possibilities (for the result) have to be taken into account. The information furnished by the outcome of the experiment reduces this uncertainty; Fisher defines the quantity of information furnished by a series of repeated measurements as the reciprocal of the variance.

Conditional and unconditional information

Information about real events that have happened (e.g. a volcanic eruption), or about entities which exist (e.g. a sequence of DNA) is primarily unconditional, i.e. it does not depend on anything. As soon as information is encoded, it becomes conditional on the code, however.

Scientific work has two stages:

1. Receiving unconditional information from nature (by making observations in the field, doing experiments in the laboratory); and

2. Generating conditional information in the form of hypotheses and theories relating the observed facts to each other using axiom systems. The success of any theory (which may be one of several) largely depends

on general acceptance of the chosen propositions and the mathematical apparatus used to manipulate the elements of the theory, i.e. there is a strongly social aspect involved.

Conditional information tends to be unified, e.g. a group of scattered tribes, or practitioners of initially disparate disciplines, may end up speaking a common language (they may then comprehend the information they exchange as being unconditional, and may ultimately end up believing that there cannot be other languages). Encoded information is conditional on agreement between emitters and receivers concerning the code.

Experiments and observations

Consider again the example of the measurement of the length of an object using a ruler and the information gained thereby. The gain presupposes the existence of a world of objects and knowledge, including the ruler itself and its calibration in units of measurement. The overall procedure is captured, albeit imperfectly, in figure 2.1.

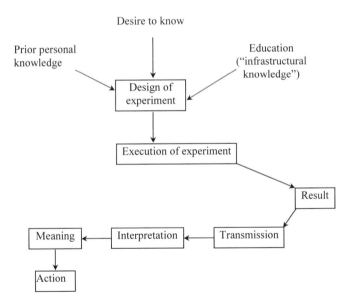

Figure 2.1: The procedures involved in carrying out an experiment, from conception to ultimate dissemination.

The essential point is that 'information' has two parts, a prior part embodied by the physical apparatus, the knowledge required to carry out the experiment or observation etc., and a posterior part equal to the loss in uncertainty about the system due to having made the observation. The prior part can be thought of as specifying the set of possible values from which the observed value must come. In a physical measurement, it is related to the structure of the experiments and the instruments it employs The posterior part (I) is sometimes called 'missing information', because once the prior part (K) is specified, the system still has the freedom, quantified by I, to adopt different microstates. In a musical analogy, K would correspond to the structure of a Bach fugue, and I to the freedom the performer has in making interpretational choices while still respecting the structure.[4] One could say that the magnitude of I corresponds to the degree of logical indeterminacy inhering in the system, i.e. that part of its description that cannot be formulated within itself; it is the amount of *selective* information lacking.

I can often be calculated according to the procedures described in the previous section, i.e. the Hartley or Shannon index. If we need to quantify K, it can be done using the concept of algorithmic information content (AIC), i.e. the length of the most concise description of what is known about the system, i.e. Kolmogorov information (see §6.5). Hence the total information[5] is the sum of the ensemble (Shannon) entropy I and the physical (Kolmogorov) entropy K:

$$\mathcal{I} = I + K \ . \tag{2.12}$$

Mackay (1950) proposed the terms 'logon' for the structural (prior) information, equivalent to K in equation (2.12), and 'metron' for the metrical (posterior) measurement. The gain in information from a measurement (equation 2.7) falls wholly within the metrical domain, of course, and within that domain, there is a prior and posterior component (cf. §5.4).

To summarize, the Kolmogorov information K can be used to define the structure of information, and is calculating by considering the system used to make a measurement. The result of the measurement is macroscopic, remembered information, quantified by the Shannon index I. The gain in information equals (final − initial information), i.e.

$$I = (I_f + K) - (I_i + K) = I_f - I_i \ . \tag{2.13}$$

In other words, it is unexceptionable to assume that the measurement procedure does not change the structural information, although this must only

---

[4]Cf. Tureck.

[5]Called the physical information of a system by Zurek.

be regarded as a cautious, provisional statement: presumably any measurement, or series of measurements, which overthrows the theoretical framework within which a measurement was made does actually lead to a change in K. Equation (2.12) formalizes the notion of quiddity *qua* essence, comprising substance (K) and properties (I). The calculation of K will be dealt with in more detail in Chapter 6. As a final remark in this section, we note that the results of an experiment or observation transmitted elsewhere may have the same effect on the recipient as if he had carried out the experiment himself.

Problem. Critically scrutinize figure 2.1 in the light of the above discussion and attempt to quantify the information flows.

2.2 Constraint

Shannon puts emphasis on the information resulting from the selection from a set of possible alternatives (implying the existence of alternatives): information can only be received where there is doubt. Much of the theory of information deals with *signals*: signals operate on the set of alternatives constituting the recipient's doubt to yield a lesser doubt, or even certainty. Thus the signals themselves have an information content by virtue of their potential for making selections; the quantity of information corresponds to the intensity of selection, or to the recipient's surprise upon receiving the information. I from equation (2.5) gives the average information content per symbol; it is a weighted mean of the degree of uncertainty (i.e. freedom of choice) in choosing a symbol before any choice is made.

If we are writing a piece of prose, and even more so if it is verse, our freedom of choice of letters is considerably circumscribed. In English, the probability that 'x' follows 'g' is much lower than $\frac{1}{26}$, or $\frac{1}{27}$, if we include, as we should, the space as a symbol. as is the probability that 'b' follows 'bb'. In other words, the selection of a particular letter depends on the preceding symbol, or group of preceding symbols. This problem in linguistics was first investigated by Markov, who encoded a poem of Pushkin's using a binary coding scheme admitting consonants (C) or vowels (V). Markov proposed that the selection of successive symbols C or V no longer depended on their probabilities as determined by their frequencies ($v = V/(V + C)$, where V and C are respectively the total numbers of vowels and consonants). To every pair of letters (L_j, L_k) there corresponds a conditional probability p_{jk}; given that L_j has occurred, the probability of L_k at the next selection is p_{jk}. If the initial letter has a probability a_j, then the probability of the sequence $(L_j, L_k, L_l) = a_j p_{jk} p_{kl}$ etc. The scheme can be conveniently written in matrix notation:

$$
\begin{array}{c|cc}
\downarrow & C & V \\
\hline
C & p_{cc} & p_{vc} \\
V & p_{cv} & p_{vv}
\end{array}
\tag{2.14}
$$

where p_{cc} means the probability that a consonant is followed by another consonant, etc. The matrix is stochastic, that is the columns must add up to one. If every column is identical, then there is no dependence on the preceding symbol, and we revert to a random, or zeroth order Markov, process. Suppose now that observation reveals that the probability of C occurring after V preceded by C is different from that of C occurring after V preceded by V, or even that the probability of C occurring after VV preceded by C is different from that of C occurring after VV preceded by V. These higher order Markov processes can be recoded in strict Markov form, thus for the second order process (dependence of the probabilities on the two preceding symbols) 'VVC' can be written as a transition from VV to VC, and hence the matrix of transition probabilities becomes

$$
\begin{array}{c|cccc}
\downarrow & CC & CV & VC & VV \\
\hline
CC & p_{ccc} & 0 & p_{cc} & 0 \\
CV & p_{ccv} & 0 & p_{vcc} & 0 \\
VC & 0 & p_{cvc} & 0 & p_{vvc} \\
VV & 0 & p_{cvv} & 0 & p_{vvv}
\end{array}
\tag{2.15}
$$

and so on for higher orders. Notice that some transitions necessarily have zero probability.

The reader may object that one rarely composes text letter by letter, but rather word by word. Clearly there are strong constraints governing the succession of words in a text. The frequencies of these successions can be obtained by counting word occurrences in very long text, and used to construct the transition matrix, which is of course gigantic even for a first order process. We may also note that a book which ends with "...in the solid state is greatly aided by this new tool." is more likely to begin with "You are working on the design of a new rocket." than one ending "I became submerged in my thoughts which sparkled with a cold light."[6]

We note here that clearly one may attempt to model DNA or protein sequences as Markov processes, as will be discussed in Part III. Markov chains as such will be discussed more fully in Chapter 6.

---

[6]Good has shown that ordinary language cannot be represented even by a Markov process of infinite order.

The notion of constraint applies whenever a set "is smaller than it might be". The classic example is of road traffic lights, which display various combinations of red, amber and green, each of which may be on or off. Although $2^3 = 8$ combinations are theoretically possible, in most countries only certain combinations are used, typically only four out of the eight. Constraints are ubiquitous in the universe and much of science consists in determining them; thus in a sense 'constraint' is synonymous with 'regularity'. Law of nature are clearly constraints, and the very existence of physical objects such as tables and aeroplanes, which have fewer degrees of freedom than their constituent parts considered separately, is a manifestation of constraint.

In this book we are particularly concerned with constraints applied to sequences. Clearly if a Markov process is in operation the variety of the set of possible sequences generated from a particular alphabet is smaller than it would be had successive symbols been freely selected, i.e. it is indeed "smaller than it might have been". 'Might have been' requires the qualification, then, of 'would have been if successive symbols had been freely (or randomly— leaving the discussion of 'randomness' to Chapter 6) selected. We already know how to calculate the entropy (or information, or Shannon index, or Shannon-Weaver index) I of a random sequence (equation 2.5); there is a precise way of calculating the entropy per symbol for a Markov process (see §6.2.1), and the reader may use the formula derived there to verify that the entropy of a Markov process is less than that of a 'perfectly random' process. Using some of the teminology already introduced, we may expand on this statement to say that the surprise occasioned by receiving a piece of information is lower if constraint is operating; for example, when spelling out a word it is practically superfluous to say 'u' after 'q'.

The constraints affecting the choice of successive words are a manifestation of the syntax of a language. In the next chapter we shall look at other ways in which constraint can operate, but for now we can simply state that whenever constraint is present, the entropy (of the set we are considering, hence of the information received by selecting a member of that set) is lower than it would be for a perfectly random selection from that set.

This maximum entropy (which, in physical systems, corresponds to the most probable arrangement, i.e. to the macroscopic state which can be arranged in the largest number of ways), let us call it $I_{\max}$, allows us to define a relative entropy I_{rel}:

$$I_{\mathrm{rel}} = \frac{\text{actual entropy}}{I_{\max}} \tag{2.16}$$

and a redundancy R:

$$R = 1 - I_{\mathrm{rel}} . \tag{2.17}$$

Most computer languages lack redundancy: a single wrong character in a program will usually cause the program to halt, or not to compile.

Problem. Identify some constraints in biology.

2.2.1 The value of information

In order to quantify value V, we need to know the goal towards which the information will be used. Two cases may be considered:

(i) The goal can almost certainly be reached by some means or another. In this case a reasonable quantification is

$$V = \text{(cost and/or time required to reach goal without the information)}$$
$$- \text{(cost and/or time required to reach goal with the information)}. \quad (2.18)$$

(ii) The probability of reaching the goal is low. Then it is more reasonable to adopt

$$V = \log_2 \frac{\text{prob. of reaching goal with the information}}{\text{prob. of reaching goal without the information}} . \quad (2.19)$$

With both these measures, irrelevant information is clearly zero-valued.

Durability of information contributes to its value. Intuitively, we have the idea that the more important the information, the longer it is preserved. In antiquity, accounts of major events such as military victories were preserved in massive stone monuments whose inscriptions can still be read today several thousand years later. Military secrets are printed on paper or photographed using silver halide film and stored in bunkers, rather than committed to magnetic media. We tend to write down things we need to remember for a long time.

The value of information is closely related to the problem of weighing the credibility that one should accord a certain received piece of information. The question of weighting scientific data from a series of measurements was an important driver for the development of probability theory. In 1777 Daniel Bernoulli raised this issue in the context of averaging astronomical data, where it was customary to simply reject data deviating too far from the mean, and weight all others equally.[7]

---

[7]D. Bernoulli, Acta Acad. Petrop., pp. 3–33.

The quality of information

Quality is an attribute which brings us back to the problem posed by D. Bernoulli in 1777, namely how to weight observations. If we return to our simple measurement of the length of a piece of wood, the reliability may be affected by the physical condition of the measuring stick, its markings, its origin (e.g. from a kindergarten or from Sèvres), the eyesight of the reader, etc.

The value of information is also related to the amount already received. One of the Bernoullis is said to have proposed that the value of an amount m of money received is proportional to $\log(m + c)/c$, where c is the amount of money already possessed, and a similar relationship may apply to information.

2.3 Accuracy, meaning and effect

2.3.1 Accuracy

In the preceding sections, we have focused on the information gained when a certain signal, or sequence of signals, is received. The quantity of this information I has been formalized according to its statistical properties. I is of particular relevance when considering how accurately a certain sequence of symbols can be transmitted. This question will be considered in more detail in Chapter 3. For now, let us merely note that no physical device can discriminate between pieces of information differing by arbitrarily small amounts. In the case of a photographic detector, for example, diminishing the difference will require larger and larger detectors in order to discriminate, but photon noise places an ultimate limitation in the way of achieving arbitrarily small detection.

A communication system depending on setting the position of a pointer on a dial to one of 6000 positions, and letting the position be observed by the distant recipient of the message through a telescope, while allowing a comfortably large range of signs to be transmitted, would be hopelessly prone to reading errors, and it was long ago realized that far more reliable communication could be achieved by using a small number of unambiguously uninterpretable signs, e.g. signalling flags at sea, which could be combined to generate complex messages.

Practical information space is thus normally discrete. For example, meteorological bulletins do not generally give the actual wind speed in km/h and the direction in degrees, but refer to one of the 13 points of the Beaufort scale and one of the eight compass points. The information space is therefore a finite 2-space with 8×13 elements.

The rule for determining the distance between two words, i.e. the metric of information space, is most conveniently perceived if the words are encoded in binary form. The Hamming distance is the number of digit places in which the two words differ.[8] This metric satisfies the usual rules for distance, i.e. if a, b and c are three points in the space and $D(a, b)$ is the distance between a and b, then

$$
\begin{aligned}
D(a, a) &= 0 \ ; \\
D(a, b) = D(b, a) &> 0 \qquad \text{if } b \neq a \ ; \\
D(a, b) + D(b, c) &\geq D(a, c) \ .
\end{aligned}
$$

In biology, the question of accuracy refers especially to the replication of DNA, its transcription into RNA, and the translation of RNA into protein. It may also refer to the accuracy with which physiological signals can be transmitted within and between cells.

2.3.2   Meaning

At the first level, Shannon's theory is deliberately divorced from the question of semantic content (i.e. meaning). In the simple example of measuring the length of a piece of wood, the question if meaning scarcely enters into the discourse. In nearly all the other cases, where we are concerned with receiving signs, or sequences of symbols, after we have received them accurately we can start to concern ourselves with the question of meaning. The issues can range from simple ones of interpretation to involved and complex ones. An example of the former is the interpretation of the order "Wait!" heard in a workshop. It may indeed mean 'pause until further notice', but heard by an apprentice standing by a weighing machine may well be interpreted as 'call out the weight of the object on the weighing pan'. An example of the latter is the statement "John Smith is departing for Paris": it has very different connotations according to whether it was made in an airport, a railway station or some other place.

It is easy to show that the meaning contained in a message depends on the set of possible messages. Ashby has constructed the following example. Suppose a prisoner-of-war is allowed to send a message to his family. In one camp, the message can be chosen from the following set:

---

[8]Cf. J.E. Surrick and L.M.Conant, *Laddergrams*, New York: Sears (1927). 'Turn bell into ring in six moves' etc.

> I am well
> I am quite well
> I am not well
> I am still alive,

and in another, only one message may be sent:

> I am well.

In both cases, there is implicitly a further alternative, no message at all, which would mean that the prisoner is dying or already dead. In the second camp, if the recipient is aware that only one message is permitted, he or she will know that it encompasses several alternatives, which are explicitly available in the first camp. Therefore, the same message (I am well) can mean different things depending on the set from which it is drawn.

In much human communication, it is the context-dependent difference between explicit and implicit meaning which is decisive in determining the ultimate outcome of the reception of information. In the latter example of the previous paragraph, the context—here provided by the physical environment—endows the statement with a large complement of implicit information, which mostly depends on the mental baggage possessed by the recipient of the information. For example, the meaning of a Chinese poem may only be understandable to someone who has assimilated Chinese history and literature since childhood, and will not as a rule be intelligible to a foreigner armed with a dictionary.

A very similar state of affairs is present in the living cell. A given sequence of DNA will have a well-defined explicit meaning in terms of the sequence of amino acids it encodes, and into which it can be translated. In the eukaryotic cell, however, the amino acid sequence may then be glycosylated and further transformed, but in a bacterium it may not be, indeed it may even misfold and aggregate—a concrete example of implicit meaning dependent on concept.

The importance of context in determining implicit meaning is even more graphically illustrated in the case of the developing multicellular organism, in which the cells are initially all identical, but according to chemical signals received from their environment will develop into different kinds of cells. The meaning of the genotype is the phenotype, and it is implicit rather than explicit meaning, which is of course why the DNA sequence of any earthly organism sent to an alien civilization will not allow them to reconstruct the organism. Ultimately most of the cells in the developing embryo become irreversibly different from each other (differentiation), but while they are still pluripotent, they may be transplanted into regions of different chemical composition and change their fate: for example, a cell from the non-neurogenic region of one embryo transplanted into the neurogenic region of another may become a neuroblast. The mechanism of such transformations will be discussed in a little more detail in Chapter 9, but here this type of phenomenon serves to illustrate how the implicit meaning of the genome dominates the explicit meaning. This implicit meaning is called epigenetics, and it seems clear that we shall not truly understand life before we have developed a powerful way of treating epigenetic phenomena. Shannon's approach has proved very powerful for treating the problem of the accurate transmission of signals, but at present we do not have a comparable foundation for treating the problem of the precise transfer of meaning.

Even at the molecular level, at which phenotype is more circumscribed and could be considered to be the function (of an enzyme), or simply the structure of a protein, there is presently little understanding of the relation between sequence and function, as illustrated by the thousands of known different sequences encoding the same type of structure and function, or different sequences encoding different structures but the same type of function, or similar structures with different function.

Part of the difficulty is that the function (i.e. biological meaning) is not so conveniently quantifiable as the information content of the sequence encoding it. Even considering the simpler problem of structure alone, there are various approaches yielding very different answers. Supposing that a certain protein has a unique structure (most nonstructural proteins have of course several structures in order to function; the best-known example is probably haemoglobin). This structure could be specified by the coordinates of all the constituent atoms, or the dihedral angles of each amino acid, listed in order of the sequence, and at a given resolution (Dewey calls this the algorithmic complexity of a protein, cf. K in equation 2.12). If, however, protein structures come from a finite number of basic types, it suffices to specify one of these types, which moves the problem back into one dealing with Shannon-type information.

In the case of function, a useful starting point could be to consider the immune system, in which the main criterion of function is the affinity of the antibody to the target antigen, or, more precisely, the affinity of a small region of the antibody. The discussion of affinity, and how affinities can lead to networks of interactions, will be dealt with in Chapter 14.

The problem of assigning meaning to a sign, or a message (a collection of signs), is usually referred to as the semantic problem. Semantic information cannot be interpreted solely at the syntactical level.

Just as a set of antibodies can be ranked in order of affinity, so may a series of statements be ranked in order of semantic precision. For example, consider the statements:

A train will leave.
A train will leave from London today.
An express train will leave from London Marylebone for Glasgow at 10 a.m. today.

and so on. Postal or e-mail addresses have a similar kind of syntactical hierarchy. Although we are not yet able to assign numerical values to meanings, we can at least order them.

Carnap and Bar-Hillel have framed a theory, rooted in Carnap's theory of inductive probability, attempting to do for semantics what Shannon did for the technical content of a message. It deals with the semantic content of declarative sentences, excluding the pragmatic aspects (dealing with the consequences or value of received information for the recipient). It does not deal with the so-called semantic problem of communication, which is concerned with the identity (or approach thereto) between the intended meaning of the sender and the interpretation of meaning by the receiver: Carnap and Bar-Hillel place this explicit involvement of sender and receiver in the realm of pragmatics.

To gain a flavour of their approach, note that the semantic content of sentence j, conditional on having heard sentence i, is content$(j|i) =$ content$(i \& j) -$ content(i), and their measure of information is defined as information$(i) = -\log_2$ content$(\text{NOT } i)$. They consider semantic noise (resulting in misinterpretation of a message, even though all its individual elements have been perfectly received), and semantic efficiency, which takes experience into account. For example, a language with the predicates W,M and C, designating warm, moderate and cold temperatures, would be efficient in a continental climate (e.g. Switzerland or Hungary), but would become inefficient with a move to the western margin of Europe, since M occurs much more frequently there.

Although the quantification of information is deliberately abstracted from the content of a message, taking content into account may allow much more dramatic compression of a message than is possible using solely the statistical redundancy (equation 2.17). Consider how words such as 'utilization' may be replaced by 'use', appellations such as 'guidance counsellor' by 'counsellor', and phrases such as 'at this moment in time' by 'at this moment'. Many documents can be reduced in length by over two thirds without any loss in meaning (but a considerable gain in readability). With simply constructed texts, algorithmic procedures that do not require the text to be interpreted can be devised. For example, all the words in the text can be counted and listed in order of frequency of occurrence, and then each sentence is assigned a score according to the numbers of the highest ranking words (apart from 'and', 'that' etc.) it contains. The sentences with the highest scores are preferentially retained.

2.3.3   Effect

A signal may be accurately received and its meaning may be understood by the recipient, but that does not guarantee that it will engender the response desired by the sender. This aspect of information deals with the ultimate result and the far-reaching consequences of a message, and how the deduced meaning is related to human purposes. The question of the value of information has already been discussed (§2.2.1), and operationally comes close to a quantification of effect.

Mackay has proposed that the quantum of effective information is that amount that enables the recipient to make one alteration to the logical pattern describing his awareness of the relevant situation, and this would appear to provide a good basis for quantifying effect. Suppose that an agent has a state of mind M_1, which comprises certain beliefs, hypotheses etc. (the prior state). The agent than hears a sentence, which causes a change to state of mind M_2, the posterior state, which stands in readiness to make a response. If the meaning of an item of information is its contribution to the agent's total state of conditional readiness for action and the planning of action (i.e. the agent's conditional repertoire of action), then the effect is the ultimate realization of that conditional readiness in terms of actual action.[9]

As soon as we introduce the notion of a conditional repertoire of action, we see that selection must be considered. Indeed, the three essential attributes of an agent are (and note the parallel with the symbolic level):

1. A repertoire, from which alternative actions can be selected;

2. An evaluator, which assigns values to different states of affairs according to either given or self-set criteria; and

3. A selector, which selects actions increasing a positive evaluation, and diminishing deleterious evaluation.

---

[9]Wiener subsumes effect into meaning in his definition of 'meaningful information'.

2.4 Further remarks on information generation

It is a moot point whether the solution of a set of equations contains more information than the equations, since the solution is implicit (and J.S. Mill insisted that induction, not deduction, is the only road to new knowledge. The exercise of intellect involves both the transformation and generation of information, the latter quite possibly involving the crossing of some kind of logical gap). Are we then no more complex than a zygote, which apparently contains all the information required to generate a functional adult?

The reception of information is equivalent to ordering, i.e. an entropy decrease, and corresponds to the various ordering phenomena seen in nature. Three categories can be distinguished:

1. Order from disorder (sometimes called "self-organization");

2. Order from order (a templating process, such as DNA replication or transcription);

3. Order from noise (microscopic information is given macroscopic expression).

The only meaningful way of interpreting the first category is to suppose that the order was implicit in the initial state, hence it is questionable whether information has actually been generated. In the second category, the volume of ordering has increased, but at the expense of more disorder elsewhere. Note that copying does not lead to an increase in the amount of information. The third category is of genuine interest, for it illuminates problems such as that of the development of the zygote, in which environmental information is given meaningful macroscopic expression.

2.5 Summary

Information is that which removes uncertainty. It has two aspects, form (what we already know about the system), and content, the result of an operation (e.g. a measurement) within the framework of our extant knowledge. Form specifies the structure of the information. This includes the specification of the set of possible messages that we can receive, or the instrument used to measure a parameter of the system. It can be quantified as the length of the shortest algorithm able to specify the system (Kolmogorov information). If we know the set from which the result of the measurement operation has to come, the (metrical) content of the operation is given by the Shannon index (reducing to the Hartley index if the choices are equiprobable). A message (e.g. a succession of symbols) that directs our selection is, upon receipt, essentially equivalent to the result of the measurement operation encoded by the message. The Shannon index assumes that the message is known with certainty, once it has been received; if it is not, the Wiener index should be used.

Information can be represented as a sign, or as a succession of signs (symbols). The information conveyed by each symbol equals the freedom in choosing the symbol. If all choices are *a priori* equiprobable, the specification of a sequence removes uncertainty maximally. In practice, there may be strong syntactical constraints imposed on the successive choices, which greatly limit the possible variety in a sequence of symbols.

In order to be considered valuable (or desired) the received information must be remembered (macroscopic information). Microinformation is not remembered. Thus, the information inherent in the positions and momenta of all the gas molecules in a room is forgotten picoseconds after its reception. It is of no value.

Information can be divided into three aspects: the signs themselves, their syntax (their relation with each other), and the accuracy with which they can be transmitted; their meaning, or semantic value, i.e. their relation to designata; and their effect (how effectively the received meaning affects the conduct of the recipient in the desired way), which may be called pragmatics, the study of signs in relation to their users, or significs, the study of significance.[10] In other words, content comprises the signs themselves, and their syntax (i.e. the relation between them); their meaning (semantic value); and their effect on the conduct of the recipient.

Meaning may be highly context-dependent. The stronger this dependence, the more implicit the meaning. If 'genotype' constitutes the signs, then 'phenotype' constitutes meaning.

The effect of receipt of information on behaviour can be quantified in terms of changes to the logical pattern describing the awareness of the recipient to his environment. In simpler terms, this may be quantified as value in terms of a change in behaviour (assuming that enough data on replicate systems or past events is available to enable the course of action that would have taken place in te absence of the information to be determined).

Information is inherently discrete (quantal) and thus based on combinatorics, which also happens to suit the spirit of the digital computer.

---

[10]The three aspects of syntactics, semantics and pragmatics are usually considered to constitute the theory of signs, or semiotics.

Chapter 3

The transmission of information

In the previous chapter, although we spoke of the recipient of a message, implying also the existence of a dispatcher, the actual process of communicating between emitter and receiver remained rather shadowy. The purpose of this chapter is to explicitly consider transmission or communication channels.

Information theory grew up within the context of the transmission of messages and did not concern itself with appraisal of the meaning of a message. Later, Shannon (and others) went on to study the redundancy present in natural languages, since if the redundancy is taken into account in coding, the message can be compressed, and more information can be sent per unit time than would otherwise be possible.

Physically, channels can be extremely varied. The archetype used to be the copper telephone wire; nowadays it would be an optical fibre. Consider the receipt of a weather forecast. A satellite orbiting the earth emits an image of a mid-Atlantic cyclone, or a remote weather station emits windspeed and temperature. Taking the first case, photons first had to fall on a detector array, initiating electrons flowing along wires. These flows were converted into binary impulses (representing black or white, i.e. light or dark, on the image) preceded by the binary address of each pixel. In turn, these electronic impulses were then converted into electromagnetic radiation and beamed earthwards, where they were converted back into electrical pulses used to drive a printer which produced an image of the cyclone on paper. This picture was viewed by the meteorologist, photons falling on his retina were converted into an internal representation of the cyclone in the meteorologist's brain, after some processing he composed some sentences expounding the meaning of the information and its likely effect, these sentences were then spoken, involving the passage of neural impulses from brain to vocal chords, the sound emitted from his mouth travelled through the air, actuating resistance, hence electronic current fluctuations in a microphone, which travelled along a wire to be again converted into electromagnetic radiation, broadcast and picked up by a wireless receiver, converted back into acoustic waves travelling through the air, picked up by the intricate mechanism of the ear, converted into nervous impulses, and processed by the brain of the listener. According to the nature of the message, muscles may then have been stimulated in order to make the listener run outside and secure objects from being blown away, etc. And perhaps during the broadcast, some words may have been rendered unintelligible by unwanted interference. It should also be mentioned that the whole process did not of course happen spontaneously, but the satellite and attendant infrastructure had previously been launched by the meteorologist with the specific purpose of providing images useful in weather forecasting.

From this little anecdote we may gather that the transmission of information involves coding and decoding (i.e. transducing) of messages, that transmission channels are highly varied physically, and that noise may degrade the information received.

Inside the cell, it may be perceived that similar processes are operating; sensors on the surface register a new carbon source, more abundant than the one on which the bacterium has been feeding, a conformational change in the sensor protein activates its enzymatic (phosphorylation) capability, some proteins in its vicinity are phosphorylated, in consequence change conformation and then bind to the promoter site for the gene of an enzyme able to metabolize the new food source; mRNA is synthesized, from which in turn the enzyme is synthesized and perhaps modified post-translationally. The protein folds to adopt a meaningful structure and enzymatic activity, and begins to metabolize the new food perfusing into the cell. Concomitant changes result in the bacterium adopting a different shape etc., i.e. its phenotype has demonstrably changed.[1]

In very general terms, semiotics is the name given to the study of signals used for communication. In the previous chapter, the issues of the accuracy of signal transmission, the syntactical constraints reducing the variety of possible signals, the meaning of signals (semantics), and their ultimate effect were broached. In this chapter we shall be mainly concerned about the technical question of transmission accuracy, although as we shall see syntactical constraints play an important rôle in considering channel capacity. We noted at the beginning of Chapter 2 that information theory has traditionally focused on the processes of transmission. In classical information theory, as exemplified by the papers of Hartley (1928) and especially Shannon (1948),

---

[1]These processes are considered in more detail in Chapter 9.

the main problem addressed is the capacity of a communication channel for error-free transmission. This problem was highly relevant to telegraph and telephone companies, who were not in the least concerned with the nature of the messages being sent over their networks

Some features involved in communication are shown in figure 3.1. There will always be a source (emitter), channel (transmission line) and sink (receiver), and encoding is necessary even in the simplest cases: For example, a man may say a sentence to a messenger who then runs off and repeats the message to the addressee, but of course to be able to do that he had to remember the sentence, which involved encoding the words as patterns of neural firing. Even if one simply speaks to an interlocutor, and regards the mouth as the source, the mouth is not the receiver: the sounds are encoded as patterns of air waves, and decoded via tiny mechanical movements in the ear.

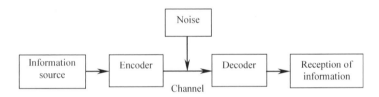

Figure 3.1: Schematic diagram of subprocesses involved in transmitting a signal from a source to a receiver. Not all the subprocesses shown are necessary, as discussed in the text. Noise may enter from the environment, or may be intrinsic to the channel hardware.

What is the flow of information in the formal scheme of figure 3.1? In the previous chapter we essentially only considered one agent, who himself carried out an operation (such as measuring the length of a piece of wood), which reduced uncertainty, and hence resulted in a gain of information according to equation (2.7), and further quantified by equation (2.5). We now consider that the information is encoded and transmitted (figure 3.2), indeed it could be broadcast to an unlimited number of people. If they desired to know the length of that piece of wood, and if the structure of their ignorance was the same as that of the measurer prior to the measurement (i.e. that the wood was less than a foot long, and they expected to receive the length in inches), then all those receiving that information would gain the same amount. The transmitted signals therefore have the potential for making a selection, by operating on the predefined set of alternatives, in exactly the same way as the actual act of measurement itself. The information content of signals is based on this potential for discrimination. Hartley, in his pioneering paper, refers to the successive selection of signs from a given list. This is of course precisely what happens when sending a telegramme.

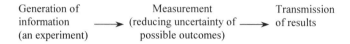

Figure 3.2: Schematic diagram of the subprocesses involved in carrying out a physical experiment and transmitting the results.

3.1   The capacity of a channel

Channel capacity is essentially dependent on the physical form of the channel. If the channel is consituted by a runner bearing a scroll on which the messsage is inscribed, the capacity, in terms of number of messages per day, depends on the distance the runner has to cover, the nature of the terrain, his physique etc.[2] The capacity of a heliograph signalling system (in flashes per minute) depends on the dexterity of the operators working the mirrors and the availability of sunlight.

It is obviously convenient, when confronted with the practicalities of comparing the capacities of different channels (for example, a general may have to decide whether to rely on runners or set up a heliograph) to have a common scale with which the capacities of different channels may be compared. A channel is essentially transmitting *variety*. A runner can clearly convey a great deal of variety, since he could bear a large number of different messages. If he can comfortably carry a sheet on which a thousand characters are written, and assuming that the characters are selected from the English alphabet plus space, then the variety of a single message is $1000 \log_2 27 = 4754$ bits to a first approximation. If the runner can convey three scrolls a day, the variety is then $3 \times 4754/(12 \times 3600) = 0.33$ bits per second, assuming 12 hours of good daylight.

---

[2]Note that here the information source is the brain of the originator of the message, and the encoder is the brain-hand-pen system which results in the message being written down on the scroll.

The heliograph operator, on the other hand, may be able to send one signal per second, with a linear variety of two (flash or no flash), i.e. during the 12 hours of good daylight he can transmit with a rate of $\log_2 2 = 1$ bit/s.

It may be, of course, that the messages the general needs to send are highly stereotyped. Perhaps there are just 100 different messages that might need to be sent. Hence they could be listed and referred to by their number in the list. Since the number 100 (in base 10) can be encoded by $\log_2 100 = 6.64$ bits, any of the hundred messages could be sent within seven seconds. And if experience showed that only ten of the messages were sent rather frequently (say with probability 0.05 each), and the remaining 90 with probability $\frac{0.5}{90}$, application of equation (2.5) shows that 5.92 bits would suffice, so that a more compact coding of the 100 messages could in principle be found.[3]

We note in passing, with reference to equation (2.12), that all the details of the physical construction of the heliograph, or whatever system is used, and including the table of 100 messages assigning a number to each one, so only the number needs to be sent, are included in K. Should it be necessary to quantify K, it can be done via the algorithmic complexity, but as far as the transmission of messages is concerned, this is not necessary, since we are only concerned with the *gain* of information by the recipient.

The meaning of each message—i.e. an encoded number—sent under the second scheme could potentially be very great. It might refer to a book full of instructions. Here we shall not consider the effect of the message (cf. 2.3.3).

Another point to consider is possible interference with the message. The runner would be a target for the enemy, hence it may be advisable to send, say, three runners in parallel with copies of the same message. It might also have been found that the distant heliograph operator had difficulty in receiving the flashes reliably from the sender, and it might therefore have been decided to repeat each flash three times and the recipient would use majority selection on each group of three to deduce the message. The capacity of the channel would be thereby lowered threefold.

In many practical cases, the physical medium for transmitting messages has to be shared by many different messages. It is a great advantage of optical communications that streams of photons of different wavelength do not interfere with one another. Therefore an optical fibre can carry many independent signals. Inside a cell, in which the cytoplasm is a shared medium, many different molecules are present and independence is determined by differential chemical affinity between pairs of molecules.

---

[3]Shannon's theory does not give any clues as to how the most compact coding can be found.

3.2 Coding

Coding refers to the transduction of a message into another form. It is ubiquitous in our world. Ideas are encoded into words, or music, or pictures etc., one language may be encoded into another, etc. We have already made extensive use of binary coding; the compact disc-based recording industry today uses binary coding almost exclusively for music, pictures and words. Evidently any number can be written in base 2, hence a possible drill for binary coding consists of the following steps:

1. assign a number to each state to be encoded;

2. convert that number into base 2.

A DNA sequence can thereby be converted into binary form by making the assignments A $\rightarrow$ 1, C $\rightarrow$ 2, T $\rightarrow$ 3, and G $\rightarrow$ 4, which in base two are 1, 10, 11, and 100 respectively. The coded sequence would have to be written (001, 010 etc.) and read in groups of three digits, otherwise 'AA' could be misinterpreted as 'T', and so on. The reading frame is thus defined as the series of groups of three beginning with the first. DNA is an example of a usually nonoverlapping code of contiguous triplets.

Codes may be written as transformations, e.g.

$$\downarrow \begin{array}{ccccc} A & B & C & D & \cdots & Z \\ B & C & D & E & \cdots & A \end{array} \quad ,$$

which could also be written down compactly by the instruction "replace each letter by the next one to the right" (sfqmbdf fbdi mfuufs cz uif ofyu pof up uif sjhiu). A scheme for recoding DNA could be:

$$\downarrow \begin{array}{cccc} A & C & T & G \\ 1 & 2 & 3 & 4 \end{array}$$

in any base above 4. As is well-known, DNA is encoded by RNA using the following transformation:[4]

$$\downarrow \begin{array}{cccc} A & C & T & G \\ U & G & A & C \end{array}$$

---

[4]Since DNA is composed of two complementary strands, one could equally well write the coding transformation as

$$\downarrow \begin{array}{cccc} A & C & T & G \\ A & C & U & G \end{array} \quad .$$

1st (5′)	second position				3rd (3′)
	U	C	A	G	
	phe	ser	tyr	cys	U
U	phe	ser	tyr	cys	C
	leu	ser	stop	stop	A
	leu	ser	stop	trp	G
	leu	pro	his	arg	U
C	leu	pro	his	arg	C
	leu	pro	gln	arg	A
	leu	pro	gln	arg	G
	ile	thr	asn	ser	U
A	ile	thr	asn	ser	C
	ile	thr	lys	arg	A
	met	thr	lys	arg	G
	val	ala	asp	gly	U
G	val	ala	asp	gly	C
	val	ala	glu	gly	A
	val	ala	glu	gly	G

Table 3.1: The genetic code.

by virtue of complementary base-pairing, and RNA triplets are in turn encoded by amino acids (table 3.2).

Codes used in telecommunications are single-valued and one-to-one transformations, i.e. bijective functions, which allows unambiguous decoding. The type of coding found in biology is more akin to that described for the broadcast meteorological bulletin, in which the physical carrier of the information changes, and the bare technical content accrues meaning. In that example, supposing that the satellite was defined as the information source, the meteorologist could scarcely have made sense, in his head, of the stream of pixel densities, but as soon as they were interpreted by writing them down as black or white squares (which he could have done with pencil on paper had he been aware of the structure of the information, especially the order in which the pixels were to be arranged) it would have been apparent that they code for a picture, i.e. there is a jump in meaning. So it is in biology: the amino acid sequence is structured in such a way that meaning is accrued, not only as a three-dimensional structure but as a functional enzyme or structural element, able to interact with other molecules.[5]

Coding—signal transduction—is ubiquitous throughout the cell, and between cells. Typically a state of a cell is encoded as a particular concentration level of a small molecule (cf. Tomkins' 'metabolic code'). For encoding this kind of information a small number of small molecules, such as cyclic adenosine monophosphate (cAMP), calcium ions (Ca^{2+}) etc. is used. The chemical nature of these molecules is usually unrelated to the nature of the information they encode (see Chapter 14 for details).

3.3   Decoding

The main requirement for decoding in a transmission scheme is that the coding transformation is one to one, and hence each encoded symbol has a unique inverse. In biological systems, decoding seems to be relatively unimportant: the encoded message is used directly, without being decoded back into its original form.

Problem.   Attempt to find an example of a decoder in a living organism.

3.4   Compression

Shannon's fundamental theorem for a noiseless channel proves that it is possible to encode the output of an information source in such a way as to transmit at an average rate equal to the channel capacity.

This is of considerable importance in telephony, which mostly deals with the transmission of natural language. Shannon himself found that the redundancy of the English language (due to syntactical constraint) is about 0.5. Hence, by suitably encoding the output of an English-speaking source, the capacity of a channel may be effectively doubled.

This compression process is well-illustrated by an example due to Shannon. Consider a source producing a sequence of letters chosen from among A, B, C and D. Our first guess would be that the three symbols were being chosen with equal probabilities of $\frac{1}{4}$, and hence the average information rate per symbol would be $\log_2 4 = 2$ bits per symbol. But suppose that after a

---

[5]Formally this seems rather mysterious. It can be thought of as a kind of noise-induced transition.

long delay we ascertain from the frequencies that the probabilities are, respectively, $\frac{1}{2}, \frac{1}{4}, \frac{1}{8}$ and $\frac{1}{8}$. Then, from equation (2.5) we determine $I = 1.75$ bits per symbol, so we should be able to encode the message (whose relative entropy is $\frac{7}{8}$ and hence redundancy R is $\frac{1}{8}$) such that a smaller channel will suffice to send it. The following code may be used:

$$\downarrow \begin{array}{cccc} A & B & C & D \\ 0 & 10 & 110 & 111 \end{array} \; .$$

The average number of binary digits used in encoding a sequence of N symbols will be $N(\frac{1}{2} \times 1 + \frac{1}{4} \times 2 + \frac{2}{8} \times 3) = \frac{7}{4}N$. 0 and 1 can be seen to have equal probabilities, hence I for the coded sequence is 1 bit/symbol, equivalent to 1.75 binary symbols per original letter. The binary sequence can be decoded by the transformation

$$\downarrow \begin{array}{cccc} 00 & 01 & 10 & 11 \\ A' & B' & C' & D' \end{array} \; .$$

The compression ratio of this process is $\frac{7}{8}$. Note, however, that there is no general method for finding the optimal coding.

Problem. Using the above coding, show that the 16 letter message 'AB-BAAADABACCDAAB' can be sent using only 14 letters.

The Shannon technique requires a long delay between receiving symbols for encoding and the actual encoding, in order to accumulate sufficiently accurate individual symbol transmission probabilities. The entire message is then encoded. This is of course a highly impractical procedure. Mandelbrot has devised a procedure whereby messages are encoded word by word. In this case the word delimiters, e.g. spaces in English text, play a crucial rôle. From Shannon's viewpoint, such a code is necessarily redundant, but on the other hand an error in a single word renders only that word unintelligible, not the whole message. It also avoids the necessity for a long delay before coding can begin.

The Mandelbrot coding scheme has interesting statistical properties. One may presume that the encoder seeks to minimize the cost of conveying a certain amount of information, using a collection of words that are at his disposal. If p_i is the probability of selecting and transmitting the ith word, then the mean information per symbol contained in the message is, as before, $-\sum p_i \log p_i$. We may suppose that the cost of transmitting a selected word is proportional to its length. If c_i is the cost of transmitting the ith word, then the average cost per word is $\sum p_i c_i$. Minimizing the distribution of the probabilities while keeping the total information constant (using Lagrange's method of undetermined multipliers) yields

$$p_i = Ce^{-Dc_i} , \qquad (3.1)$$

a sort of Boltzmann distribution. C is a constant fixed by the condition that $\sum p_i = 1$, and D is an as yet undetermined constant.

Suppose that the words are made up of individual letters (symbols), and demarcated by a special word demarcation symbol (the space in many languages). Cost, length and number of letters are all proportional to each other. If the letters can be chosen in any way from an alphabet of A different ones, by the multiplication rule (§4.2.1) there are A^n different n-letter words. Let these words now be ranked in order of increasing cost and call this rank r. Since the cost increases linearly with n, it only increases logarithmically with rank,[6] i.e.

$$c_r = \log_A r . \qquad (3.2)$$

Substituting (3.2) into (3.1), one obtains a power law relation

$$p_r = Cr^{-B} \qquad (3.3)$$

known as Zipf's law when $B = 1$. Mandelbrot has shown that, more precisely, (3.3) is

$$p_r = C(r + \rho)^{-B} \qquad (3.4)$$

and that the constant B (subsuming D in 3.1), the reciprocal of the informational temperature θ of the distribution (by analogy with the thermodynamic case), can take values other than 1: for $B > 1$, i.e. $\theta < 1$, the language is called open (because the value of C does not greatly depend on the total number of words), whereas for $B < 1$ it does, and the corresponding language is called closed. The constant ρ is connected with the freedom of choosing words (cf. §4.2.3), but a deep interpretation of its significance in messages has not yet been given. Equation (3.4) fits the distribution of written texts remarkably well, and most languages such as English, German etc. are open, whereas highly stylized languages (e.g. modern Hebrew, the English of the Pennsylvania Dutch) are closed. θ is a measure of the agility of exploiting vocabulary; low values are characteristic of children learning a language, or schizophrenic adults; the richest and most imaginative use of vocabulary corresponds to $\theta = 1$.

A useful way of compressing long sequences of symbols is to search for segments that are duplicated. The duplicates can then be encoded by the

---

[6]The words are listed in order of increasing cost; rank 1 has the lowest cost, and so on.

distance of the match from the original sequence and the length of the matching sequence (number of symbols). Zipping software typically works on this principle; the compression is greatest for files with a lot of repetitive material, but according to van der Waerden's extension of Baudet's conjecture any string of two kinds of symbols has repetitive sequences of at least one of the symbols.

Suppose two ergodic binary sources P and Q emit ones with probabilities p and q. The Kullback-Leibler relative entropy (in bits per character) between the two strings is

$$S_{\text{PQ}} = -q \log_2 \frac{p}{q} - (1 - q) \log_2 \frac{1 - p}{1 - q} \qquad (3.5)$$

and may be used as a measure of distance between the two strings. Benedetto, Caglioti and Loreto (unpublished) have devised an ingenious method for determining S_{PQ} from two sources by zipping a long string from each source (P and Q), and the same long strings to each of which are appended a short string (say P') from one of the sources. S_{PQ} is then the difference in coding efficiency between P' coded optimally because it follows P (the source is ergodic) and P' coded nonoptimally because it follows Q. If L is the length of the zipped file, then

$$S_{\text{PQ}} = [L_{P+P'} - L_P - (-L_{Q+P'} - Q)]/L'_{P'} \qquad (3.6)$$

where $L'_{P'}$ is the (unzipped) length of the original short string.

Ergodicity

Ergodicity means that every allowable point in phase space is visited infinitely often in infinite time, or, in practice, every allowable point in phase space is approached arbitrarily closely after a long time. It is of course a pillar of Boltzmann's assumption that the microstates of an ensemble have equal *a priori* probabilities, and indeed of the rest of statistical mechanics. Nevertheless, as our knowledge of the world has increased it has become apparent that ergodicity actually applies only to a small minority of natural systems. Although some systems may not even be ergodic in the infinite time limit, most observed departures from ergodicity occur because of the inordinately long times that would be required to satisfy it. The departures are particularly common in condensed matter—any glass, for example, exhibits broken ergodicity. In nonergodic systems, the phase space or ensemble average does not equal the time average.

A homely illustration of some of the issues to be considered, in particular that breaking depends on the timescale of the observer, is provided by a cup of hot coffee to which cream is added and stirred. The coffee and cream become homogeneously mixed after a minute or so, the cup and contents reach the temperature of the surroundings after tens of minutes, and the water evaporates and is in equilibrium with the atmosphere in the room after a few thousand minutes. Whether the observed behaviour is representative of the allowed phase space depends on the observational timescale τ_0. In general, broken ergodicity can be expected if there are significant dynamical timescales longer than τ_0.

3.5   Noise

So far we have supposed that the messages received over the communication channel are precisely those transmitted. This is a rather idealized situation. We have doubtless had the experience of speaking on a very noisy telephone line, or listening to a radio wth very poor reception, and only been able to make out one word in two perhaps, and yet could still understand what was being said. The syntactical redundancy of English is about 0.5, hence it is not surprising that about half the words or symbols may be removed (at random) without overly impairing our ability to receive the original message.

According to our previous discussion of the Shannon index, I is additive for independent sources of uncertainty. Noise is an independent source of uncertainty, and can be treated within the theoretical framework we have discussed.

Suppose that signal x was sent, and y was received, the difference between the two being due to noise. We shall call the amount of information lost in transmission the equivocation E.

Definition.   The equivocation is

$$E = I(x) - I(y) + I_x(y) \tag{3.7}$$

where $I(x)$ is the information sent, $I(y)$ the information received, and $I_x(y)$ the uncertainty in what was received if the signal sent be known.[7]

The concept of equivocation enables one to write the actual rate of information transmission $\mathcal{R}$ over a noisy channel in a rather transparent way:

$$\mathcal{R} = I(x) - E , \tag{3.8}$$

---

[7]It should be clear that the information sent is already the result of some measurement operation or whatever, in the sense of our previous discussion.

i.e. the rate equals the rate of transmission of the original signal minus the
uncertainty in what was sent when the message received is known. From our
definition (3.7),

$$\mathcal{R} = I(y) - I_x(y) \,, \tag{3.9}$$

where $I_x(y)$ is the spurious part of the information received, i.e. the part due
to noise, or, equivalently, the average uncertainty in a message received when
the signal sent is known. It follows (§4.1) that

$$\mathcal{R} = I(x) + I(y) - I(x, y) \,, \tag{3.10}$$

where $I(x, y)$ is the joint entropy of input (information transmitted) and
output (information received). By symmetry, the joint entropy equals

$$I(x, y) = I(x) - I_x(y) = I(y) - I_y(x) \,. \tag{3.11}$$

We could just as well write E as $I_y(x)$, i.e. it is the uncertainty in what was
sent when it is known what was received. If there is no noise, $I(y) = I(x)$
and $E = 0$.

Let the error rate be η per symbol. Then

$$E = I_y(x) = \eta \log \eta + (1 - \eta) \log(1 - \eta) \,. \tag{3.12}$$

The maximum error rate is 0.5 for a binary transmission; the equivocation
is then 1 bit/symbol, and the rate of information transmission is zero.

The equivocation is just the conditional or relative entropy, and can also
be derived using conditional probabilities. Let $p(i)$ be the probability of the
ith symbol being transmitted, and $p(j)$ be the probability of the jth symbol
being received.

$p(j|i)$ is the conditional probability of the jth signal being received when
the ith was transmitted, $p(i|j)$ is the conditional probability of the ith signal
being transmitted when the jth was received (posterior probability), and
$p(i, j)$ is the joint probability of the ith signal being transmitted and the jth
received.

The ignorance removed by the arrival on one symbol is (cf. equation 2.7)

$$
\begin{aligned}
I &= \text{initial uncertainty} - \text{final uncertainty} \\
&= \log p(i) - (-\log p(j)) \\
&= \log \frac{p(i|j)}{p(i)} \,.
\end{aligned} \tag{3.13}
$$

Averaging over all i, j:

$$\bar{I} = \sum_i \sum_j p(i, j) \log \frac{p(i|j)}{p(i)} \tag{3.14}$$

but since $p(i,j) = p(i)p(j|i) = p(j)p(i|j)$ (cf. §5.2.2),

$$\bar{I} = \sum_i \sum_j p(i,j) \log \frac{p(i,j)}{p(i)p(j)} \ . \tag{3.15}$$

If $i = j$ always, then we recover the Shannon index, equation (2.5). If the two are statistically independent, $\bar{I} = 0$.

From our definition of $p(i,j)$ we can write the posterior probability as

$$p(i,j) = \frac{p(i)}{p(j)} p(j,i) \ . \tag{3.16}$$

Shannon's fundamental theorem for a discrete channel with noise proves that, if the channel capacity be $\mathcal{C}$ and the source transmission rate be $\mathcal{R}$, then if $\mathcal{R} \leq \mathcal{C}$ there exists a coding system such that the source output can be transmitted through the channel with an arbitrarily small frequency of errors. The capacity of a noisy channel is defined as

$$\mathcal{C}_{\text{noisy}} = \max(I(x) - E) \ , \tag{3.17}$$

the maximization being over all sources that might be used as input to the channel.

3.6   Error correction

Suppose a binary transmission channel had a 20% chance of transmitting an incorrect signal, hence a message sent as '0110101110' might appear as '1100101110'. An easy way to render the system immune from such noise would be to repeat each signal threefold and incorporate a majority detector in the receiver. Hence the signal would be sent as '000111111000111000111111-111000' and received as '001011011000110000101111111100' (say), but majority detection would still enable the signal to be corectly restored. The penalty, of course, is that the channel capacity is reduced to a third of its previous value.

Many physical devices are so designed to be immune, to a certain degree, to random fluctuations in the physical quantities encoding information. In a digital device, zero voltage applied to a terminal represents the digit '0', and 1 V (say) represents the digit '1'. In practice, any voltage up to about 0.5 will be interpreted as zero, and all voltages above 0.5 will be interpreted as 1.0 (see figure 3.3).

It is perfectly possible to devise codes that can detect and correct errors. Hamming defines systematic codes as those in which each code symbol has

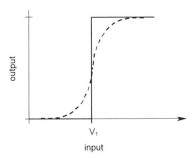

Figure 3.3: Output-input relationships for a device such as an electrome-chanical relay (solid line) and a field-effect transistor (dashed line).

exactly n binary digits, m being associated with the information being conveyed and $k = n - m$ being used for error detection and correction. The redundancy (cf. equation 2.17 of a systematic code is defined as

$$R_{s.c.} = n/m \ . \tag{3.18}$$

Hamming constructed a single error detecting code as follows: information is placed in the first $n-1$ positions of n binary digits. Either a zero or a one is placed in the nth position, the choice being made to ensure an even number of ones in the n digit word. A single (or odd number of) error would leave an odd number of ones in the word. Clearly the redundancy is $n/(n-1)$. This type of error detecting code is called a parity check; this particular one is an even parity check. n should be small enough such that the probability of more than one error is negligible.

To make an error correcting code, a larger number $(k > 1)$ of positions are given to parity checking, and filled with values appropriate to selected information positions. When the message is received, k checks are applied in order, and if the observed value agrees with the previously calculated value one writes a 0, but a 1 if it disagrees, in a new number called the checking number, which must give the position of any single error, i.e. it must describe $m + k + 1$ different things, hence k must satisfy

$$m + k + 1 \le 2^k \le 2^n/(n+1) \ . \tag{3.19}$$

The principle obviously can be extended to double error correcting codes, which of course further increases the redundancy.

3.7   Summary

Messages may be encoded in order to send them along a communication channel. Shannon's fundamental theorem proves that a message with redundancy can always be encoded to take advantage of it, i.e. a channel can transmit information up to its maximum capacity.

The capacity of a channel is the number of symbols m that can be transmitted in unit time multiplied by the average information per symbol,

$$\mathcal{C} = m\bar{I} \ . \tag{3.20}$$

Any strategy for compressing a message is actually a search for regularities in the message, and thus compresssion of transmitted information actually lies at the heart of general scientific endeavour.

Noise added to a transmission introduces equivocation, but it is possible to transmit information through a noisy channel with an arbitrarily small probability of error, at the cost of lowering the channel capacity. This introduces redundancy, defined as the quotient of the actual number of bits to the minimum number of bits necessary to convey the information. Redundancy therefore opposes equivocation, i.e. it enables noise to be overcome. Many natural languages have considerable redundancy. Technical redundancy arises through syntactical constraints. The degree of semantic redundancy of English, or any other language, is currently unknown.

Problem.   Attempt to define, operationally or otherwise, the terms message, message content, message structure.

Problem.   Calculate the amount of information in a string of DNA coding for a protein. Repeat for the corresponding messenger RNA and amino acid sequences. Is the latter the same as the binformation contained in the final folded protein molecule?

Problem.   Discuss approaches to the problem of determining the minimum quantity of information necessary to encode the specification of an organ.

Chapter 4

Sets and combinatorics

4.1 The notion of set

Set is a fundamental, abstract notion. A set is defined as a collection of objects, which are called the *elements* or *points* of the set. The notions of union ($A \cup B$, where A and B are each sets), intersection ($A \cap B$) and complement (A^c) correspond to everyday usage. Thus, if $A = \{a, b\}$ and $B = \{b, c\}$, $A \cup B = \{a, b, c\}$, $A \cap B = \{b\}$) and $A^c = \{c, d, \ldots, z\}$ if our world is the English alphabet. *Functions* can be thought of as operations which map one set onto another.

Typically, all the elements of a set are of the same type. For example, a set called 'apples' may contain apples of many different varieties, differing in their colours and sizes, but no oranges or mangoes. A set called 'fruit' could, however, contain all these, but no meat or cheese, etc.

One is often presented with the problem of finding or estimating the size of sets. Size is the most basic attribute, even more basic than types of elements. If the set is small, the elements can be counted directly, but this quickly becomes tedious, and, as the set becomes large, it is usually unnecessary to know the exact size. Hence computational short-cuts have been developed. These short-cuts are usually labelled combinatorics. Combinatorial problems are often solved by looking at them in just the right way, and at an advanced level, problems tend to be solved by clever tricks rather than the application of general principles.

Problem. Draw Venn diagrams corresponding to $\cap$, $\cup$ and complement.

4.2   Combinatorics

Most counting problems can be cast in the form of making selections, of which there are four basic types, corresponding to with or without replacement, each with or without ordering. This is equivalent to assembling a collection of balls by taking them from boxes containing different kinds of balls.

The basic law of multiplication

Consider an ordered n-tuple $(a_1, \ldots, a_n)$, in which each member a_i belongs to a set with k_i elements. The total number of possible selections equals $k_1 k_2 \ldots k_n$. For example, we select n balls, one from each of n boxes, where the ith box contains k_i different balls.

4.2.1   Ordered sampling with replacement

If all the sets from which successive selections are taken are the same size n, the total number of ordered selections of r objects from n with repetition (replacement) allowed follows from the multiplication rule

$$\prod_i^r n_i = n^r . \tag{4.1}$$

In terms of putting balls in a row of cells, this is equivalent to filling r consecutive cells with n possible choices of balls for each one; after taking a ball from a central reservoir, it is replenished with an identical ball.

4.2.2   Ordered sampling without replacement

If the balls are not replenished after removal, there are only $(n-1)$ choices of ball for filling the second cell, $(n-2)$ for the third, and so on. If the number of cells equals the number of balls, i.e. $r = n$, then there are $n!$ different arrangements—this is called a permutation (and can be thought of as a bijective mapping of a set onto itself); more generally, if $r \leq n$, the number of arrangements is

$$^nP_r = n(n-1)\ldots(n-r+1) = \frac{n!}{(n-r)!} , \tag{4.2}$$

remembering that 0! is defined as being equal to 1.

Random choice

This means that all choices are equally probable. For random samples of fixed size, all possible samples have the same probability n^{-r} with replacement, and $1/^nP_r$ without replacement. The probability of no repetition in a sample is therefore given by the ratio of these probabilities, i.e. $^nP_r/n^r$. Critera for randomness are dealt with in detail in Chapter 6.

Stirling's formula

This is useful for (remarkably accurate) approximations to $n!$, even for small values of n:

$$n! \sim (2\pi)^{\frac{1}{2}} n^{(n+\frac{1}{2})} e^{-n} \ . \tag{4.3}$$

A simpler, less accurate but easier to remember formula is

$$\log n! \sim n \log n - n \ . \tag{4.4}$$

4.2.3 Unordered sampling without replacement

Suppose now that we repeat the operation carried out in the previous subsection, but without regard to the order, i.e. we simply select r elements from a total of n. We call the number of ways it can be done W. After having made the selection, we then order the elements, to arrive at the result of the previous subsection; i.e. each selection can be permuted in $r!$ different ways. These two operations give us the following equation:

$$\frac{n!}{(n-r)!} = Wr! \tag{4.5}$$

The expression for W, which we shall now write as ^nC_r (the number of combinations of r objects out of n), or as $\binom{n}{r}$, follows immediately:

$$^nC_r = \binom{n}{r} = \frac{n!}{r!(n-r)!} \tag{4.6}$$

with $\binom{n}{0} = 1$ from the definition of $0! = 1$. This is equivalent to stating that a population of n elements has $\binom{n}{r}$ different subpopulations of size $r \leq n$. Note that

$$\binom{n}{r} = \binom{n}{n-r} \quad \text{for } r = 0, 1, ..., n \ ; \tag{4.7}$$

in words, selecting 5 objects out of 9 is the same as selecting 4 to be omitted, for example.

It is implied that the selections are independent. In practical problems, this may be far from reality. For example, a manufacturer assembling engines from 500 parts may have to choose from a total of 9000. The number of combinations is at first sight a huge number, $9000!/(500!\,8500!) \sim 10^{840}$ by Stirling's approximation, posing a horrendous logistics problem. Yet many of the choices will fix others; strong constraints will drastically reduce the freedom of choice of components.

Partitioning

The number of ways in which n elements can be partitioned into k sub-populations, the first containing r_1 elements, the second r_2 etc., where $r_1 + r_2 + \ldots + r_k = n$ is given by multinomial coefficients $n!/(r_1!r_2!\ldots r_k!)$, obtained by repeated application of (4.6). If r balls are placed in n cells with occupancy numbers $r_1, r_2, \ldots, r_n$, with all n^r possible placements equally possible, then the probability to obtain a set of given occupancy numbers equals $n^{-r}n!/(r_1!r_2!\ldots r_k!)$ (the Maxwell-Boltzmann distribution). This multinomial coefficient will be denoted using square brackets, i.e.

$$\begin{bmatrix} r \\ r_i \end{bmatrix} = \frac{n!}{r_1!r_2!\ldots r_k!} \text{ , with } \sum_{i}^{i=k} r_i = n \text{ .} \tag{4.8}$$

Fermi-Dirac statistics

Fermi-Dirac statistics are based on the following hypotheses: (i) no more than one particle can be in any given cell (hence $r \leq n$); and (ii) all distinguishable arrangements satisfying (i) have equal probabilities.

By virtue of (i), an arrangement is completely specified by stating which of the n cells contain a particle; since there are r particles, the filled cells can be chosen in $\binom{n}{r}$ ways, each with probability $\binom{n}{r}^{-1}$.

Bose-Einstein statistics

Let the occupancy numbers of the cells be given by

$$r_1 + r_2 + \cdots + r_n = r \text{ .} \tag{4.9}$$

The number of distinguishable distributions (if the balls are indistinguishable, distributions are distinguishable only if the corresponding n-tuples $(r_1, \ldots, r_n)$ are not identical) is the number of different solutions of eqn 4.9. We call this $A_{r,n}$ (given by eqn 4.11 below) and each solution has the probability $A_{r,n}^{-1}$ of occurring.

Problem. Consider a sequence of two kinds of elements, a alphas, numbered 1 to a, and b betas numbered $a+1$ to $a+b$. Show that the alphas and betas can be arranged in exactly

$$\frac{(a+b)!}{a!b!} = \binom{a+b}{a} = \binom{a+b}{b}$$

distinguishable ways.

4.2.4 Unordered sampling with replacement

This last of the four basic selection possibilities is exemplified by throwing r dice, i.e. placing r balls into $n = 6$ cells. The event is completely described by the occupancy numbers of the cells. For example, 3,1,0,0,0,4 represents three ones, one two, and three sixes.

Generalizing, every n-tuple of integers satifsying

$$r_1 + r_2 + \ldots + r_n = r \tag{4.10}$$

describes a possible configuration of occupancy numbers. Let the n cells be represented by the n spaces between $n+1$ bars. Let each object in a cell be represented by a star (for the example given above, the representation would be $| \ast\ast\ast \,|\, \ast \,||||\, \ast\ast\ast\ast \,|$). The sequence of stars and bars starts and ends with a bar, but the remaining $n-1$ bars and the r elements placed in the cells can appear in any order. Hence the number of distinguishable distributions $A_{r,n}$ equals the number of ways of selecting r places out of $n-1+r$ symbols. From equation (4.6) this is

$$A_{r,n} = \binom{n-1+r}{r} = \binom{n-1+r}{n-1} . \tag{4.11}$$

If we impose a condition that no cell be empty, the r stars leave $r-1$ spaces of which $n-1$ are to be occupied by bars, hence there are $\binom{r-1}{n-1}$ choices.

Problem. How many different DNA hexamers are there? How many different hexapeptides?

4.3 The binomial theorem

Newton's binomial formula:

$$(a+b)^n = \sum_{k=0}^{n} \binom{n}{k} a^k b^{n-k} , \tag{4.12}$$

where a and b can also be compound expressions, can be derived by combinatorial reasoning. For example, $(a+b)^5 = (a+b)(a+b)(a+b)(a+b)(a+b)$, and to generate the terms an a or b is chosen from each of the five factors.

Problem.   Generalize the binomial theorem by replacing the binomial $a+b$ by a multinomial $a_1 + a_2 + \ldots + a_r$.

Chapter 5

Probability and likelihood

5.1 The notion of probability

In everyday speech, statements such as "probably the train will be late", "probably it will be foggy tomorrow" etc. have the character of judgements. Formally, however (i.e. in the sense used throughout this book), probabilities do not refer to judgments, but to *possible results (outcomes) of an experiment*. These outcomes constitute the "sample space".[1] For example, attributing a probability of 0.6 to an event means that the event is expected to occur 60 times out of 100. This is the "frequentist" concept of probability, based on *random choices from a defined population*.

The frequentist concept is sometimes called the "objective" school of thought: the probability of an event is regarded as an objective property of the event (which has occurred), measurable via the frequency ratios in an actual experiment. Historically it has been opposed by the "subjective" school,[2] which regards probabilities as expressions of human ignorance; the probability of an event merely formalizes the feeling that an event will occur, based on whatever information is available. The purpose of theory is then merely to help in reaching a plausible conclusion when there is not enough information to enable a certain conclusion to be reached. A pillar of this school is Laplace's Principle of Insufficient Reason: two events are to be assigned equal probabilities if there is no reason to think otherwise. Under such circumstances, if information were really lacking, the objectivist would refrain from attempting to assign a probability.

These differing schools have a bearing on the whole concept of causality,

---

[1] Called *Merkmalraum* ('label space') in R. von Mises' (1931) treatise *Wahrscheinlichkeitsrechnung*.

[2] Its protagonists include Laplace, Keynes and Jeffreys.

and it may be useful to recall here some remarks of Max Planck.[3] One starts with the proposition that a necessary condition for an an event to be causally conditioned is that it can be predicted with certainty. If, however, we compare a prediction of a physical phenomenon with more and more accurate measurements of that phenomenon, one is forced to reach a remarkable conclusion, that in not a single instance is it possible to predict a physical event exactly, unlike a purely mathematical calculation. The "indeterminists" interpret this state of affairs by abandoning strict causality, and asserting that every physical law is of a statistical nature; the opposing school asserts that the laws of nature apply to an idealized world-picture, in which phenomena are represented by precise mathematical symbols, which can be operated on according to strict and generally agreed rules, and to which precise numbers can be assigned (to which an actual measurement can only approximate). In the mentally constructed world-picture all events follow certain definable laws and are strictly determined causally; the uncertainty in the prediction of an event in the world of sense is due to the uncertainty in the translation of the event from the world of sense to the world-picture and vice versa. It is left to the interested reader to pursue the implications with respect to quantum mechanics (with which we shall not be explicitly concerned in this book).

Before any further discussion about probability can take place, it is essential to agree on what is meant by the *possible results from an experiment (or observation)*. These results are called 'events'. Very often abstract models, corresponding to idealized events, are constructed to assist in the analysis of a phenomenon.

5.2 Fundamentals

The elementary unit in probability theory is the *event*. One has a fair freedom to define the event; simple events are irreducible, and compound events are combinations of simple events. For example, the throw of a die to produce a five (with probability 1/6) is a simple event, and combinations of events to yield the same final result, such as three twos, or a five and a one, are compound events. Implicitly, the level of description is fixed when speaking of events in this way: clearly the "event" of throwing a six requires many 'subevents' (which are events in their own right) involving muscular movements and nervous impulses, but these take place on a different level.

The general approach to solving a problem requiring probability is:

---

[3]Made during the 17th Guthrie Lecture to the Physical Society in London.

1. Choose a set to represent the possible outcomes;

2. Allocate probabilities to these possible outcomes.

The results of probability theory can be derived from three basic axioms, referring to events and their totality in a manner which we must take to be carefully circumscribed:[4]

$$P\{E\} \geq 0 \text{ for every event } E , \tag{5.1}$$

$$P\{S\} = 1 \text{ for the certain event } S , \tag{5.2}$$

$$P\{A\} = \sum_i P\{a_i\} . \tag{5.3}$$

S includes all possible outcomes. Hence, if E and F are mutually exclusive events, the probability of their joint occurrence (corresponding to the AND relation in logic, i.e. 'E and F') is simply the sum of their probabilities:

$$P\{E \cup F\} = P\{E\} + P\{F\} . \tag{5.4}$$

Simple events are by definition mutually exclusive ($P\{E\} \cap P\{F\} = 0$), but compound events may include some simple events that belong to other compound events, and more generally (inclusive OR, i.e. 'E or F or both'),

$$P\{E \cup F\} = P\{E\} + P\{F\} - P\{EF\} . \tag{5.5}$$

If events are independent, then the probability of occurrence of those portions shared by both is

$$P\{E \cap F\} = P\{EF\} = P\{E\}P\{F\} . \tag{5.6}$$

It follows that for equally likely outcomes (such as the possible results from throwing a die, or selecting from a pack of cards), the probabilities of compound events are proportional to the numbers of equally probable simple events which they contain:

$$P\{A\} = \frac{N\{A\}}{N\{S\}} . \tag{5.7}$$

We used this result at the beginning of this section to deduce that the probability of obtaining a five from the throw of a die is $1/6$.

---

[4]Notation. In this chapter $P\{X\}$ denotes the probability of event X; $N\{X\}$ is the number of simple events in (compound) event X. S denotes the certain event which contains all possible events. Sample space and events are primitive (undefined) notions (cf. line and point in geometry).

Problem.   Prove equations (5.4) and (5.5) with the help of Venn diagrams.

5.2.1   Generalized union

The event that at least one of N events $A_1, A_2, \ldots, A_N$ occurs, i.e. $A = A_1 \cup A_2 \cup \cdots \cup A_N$, needs information not only about the individual events but about all possible overlaps.

Theorem.   The probability P_1 of the realization of at least one among the events $A_1, A_2, \ldots, A_N$ is given by

$$P_1 = S_1 - S_2 + S_3 - S_4 + \ldots \pm S_N . \tag{5.8}$$

where the S_r are defined as the sums of all probabilities with r subscripts, e.g. $S_1 = \sum p_i$, $S_2 = \sum p_{ij}$ and $i < j < k < \cdots \le N$ so that each contribution appears only once, hence each sum S_r has $\binom{N}{r}$ terms, and the last term S_N gives the probability of the simultaneous realization of all terms.[5]
     This result can be used to solve an old problem. Consider two sequences of N unique symbols, which differ only in the order of occurrence of the symbols, which are then compared, symbol by symbol. What is the probability P_1 that there is at least one match? Let A_k be the event that a match occurs at the kth position. Therefore symbol number k is at the kth place, and the remaining $N - 1$ are anywhere. Therefore

$$p_k = \frac{(N-1)!}{N!} = \frac{1}{N} ,$$

and for every combination i, j

$$p_{ij} = \frac{(N-2)!}{N!} = \frac{1}{N(N-1)} .$$

Each term in S_r in (5.8) equals $(N-r)!/N!$ and therefore $1/r!$, therefore

$$P_1 = 1 - \frac{1}{2!} + \frac{1}{3!} - + \cdots \pm \frac{1}{N!} . \tag{5.9}$$

One might recognize that $1 - P_1$ represents the first $N + 1$ terms in the expansion of $1/e$, hence $P_1 \approx 1 - 1/e \approx 0.632$. It seems rather remarkable that P_1 is independent of N. For problems of matching genes etc. it is useful to consider an extension, that for any integer $1 \le m \le N$ the probability

---

[5]The proof is given in Feller, Chapter IV.

$P_{[m]}$ that exactly m among the N events $A_1, \ldots, A_N$ occur simultaneously is[0]

$$P_{[m]} = S_m - \binom{m+1}{m}S_{m+1} + \binom{m+2}{m}S_{m+2} - + \cdots \pm \binom{N}{m}S_N \quad (5.10)$$

and

$$P_{[0]} = 1 - P_1 = 1 - 1 + \frac{1}{2!} - \frac{1}{3!} + - \cdots \pm \frac{1}{(N-2)!} \mp \frac{1}{(N-1)!} \pm \frac{1}{N!}$$

$$P_{[1]} = 1 - 1 + \frac{1}{2!} - \frac{1}{3!} + - \cdots \pm \frac{1}{(N-2)!} \mp \frac{1}{(N-1)!}$$

$$P_{[2]} = \frac{1}{2!}\left[1 - 1 + \frac{1}{2!} - \frac{1}{3!} + - \cdots \pm \frac{1}{(N-3)!} \mp \frac{1}{(N-2)!}\right]$$

$$\vdots$$

$$P_{[N-1]} = \frac{1}{(N-1)!}\{1 - 1\} = 0$$

$$P_{[N]} = \frac{1}{N!} \, .$$

Noticing again the similarity with the expansion of $1/e$, for large N

$$P_{[m]} \approx \frac{e^{-1}}{m!} \, , \quad (5.11)$$

i.e. a special case of the Poisson distribution with $\lambda = 1$.

The probability P_m that m or more (i.e. at least M) of the events $A_1, A_2,$ $\ldots, A_N$ is

$$P_m = P_{[m]} + P_{[m+1]} + \cdots P_{[N]} \, . \quad (5.12)$$

Starting with (5.9) and noting that

$$P_{[m+1]} = P_m - P_{[m]} \, , \quad (5.13)$$

by induction, for $m \geq 1$,

$$P_{[m]} = S_m - \binom{m}{m-1}S_{m+1} + \binom{m+1}{m-1}S_{m+2} - \binom{m+2}{m-1}S_{m+3} + \cdots \pm \binom{N-1}{m-1}S_N.$$

$$(5.14)$$

5.2.2   Conditional probability

The notion of *conditional probability* is of great importance.[1] It refers to questions of the type, "what is the probability of event A, given that H has

---

[1]Indeed, Reichenbach, Popper and others have taken the view that conditional probability may and should be chosen as the basic concept of probability theory. We might in any case note that most of the results derived for (unconditional) probabilities are also valid for conditional probabilities.

occurred?" We use the notation $P\{A|H\}$ (read as "the conditional probabil-
ity of A on hypothesis H", or "the conditional probability of A for a given
event H"), and

$$P\{A|H\} = \frac{P\{AH\}}{P\{H\}} \ . \tag{5.15}$$

This result can be derived by noting that we are asking "to what extent is
H contained in A?", which means "to what extent are H and A likely to
occur simultaneously?" In set notation, this is $P\{A \cap H\} = P\{H \cap A\}$.
Therefore $P\{A|H\} = kP\{A \cap H\}$, where k is a constant. If $A = H$, then
$P\{H|H\} = kP\{H \cap H\} = kP\{H\} = 1$, hence $k = 1/P\{H\}$, and we obtain

$$P\{A|H\} = \frac{P\{A \cap H\}}{P\{H\}} \ , \tag{5.16}$$

i.e. equation (5.15). If all sample points have equal probabilities, then

$$P\{A|H\} = \frac{N\{AH\}}{N\{H\}} \ , \tag{5.17}$$

where $N\{AH\}$ is the number of sample points common to A and H.
    From this comes a theorem, due to Bayes, of great importance and widely
referred to, which gives the probability that the event A, which has occurred,
is the result of the cause E_k:

$$P\{E_k|A\} = \frac{P\{A|E_k\}P\{E_k\}}{\sum_{j=1}^{n} P\{A|E_j\}P\{E_j\}} \qquad \text{for } k = 1, ..., n \tag{5.18}$$

where the E_j are mutually exclusive hypotheses.

Proof. Let the simple events E_i be labelled such that

$$A = E_1 \cup E_2 \cup ... \cup E_m \ , \ 1 \le m \le n \ . \tag{5.19}$$

Then

$$P\{A\} = \sum_{j=1}^{m} P\{E_j\} \ . \tag{5.20}$$

From the definition (5.15),

$$\sum_{j=1}^{n} P\{A|E_j\}P\{E_j\} = \sum_{j=1}^{n} P\{A \cap E_j\} \tag{5.21}$$

But the right hand side of (5.20)

$$\sum_{j=1}^{n} P\{A \cap E_j\} = \sum_{j=1}^{m} P\{E_j\} = P\{A\} , \qquad (5.22)$$

This result can be used to write the denominator of the right hand of (5.18) as $P\{A|E_k\}P\{E_k\}/P\{A\}$, but this, according to (5.16) and after cancelling equals $P\{A \cap E_k\}/P\{A\} = P\{E_k \cap A\}/P\{A\}$, which, again using (5.16), equals $P\{E_k|A\}$. QED.

5.2.3 Bernoulli trials

Bernoulli trials are defined as repeated, (stochastically) independent trials[2] (hence probabilities multiply) with only two possible outcomes per trial, success (s) or failure (f), with respective constant (throughout the sequence of trials) probabilities p and $q = 1 - p$. The sample space of each trial is $\{s, f\}$, and the sample space of n trials contains 2^n points. The event "k successes, with $k = 0, 1, ..., n$, and $n - k$ failures in n trials" can occur in as many ways as k letters can be distributed among n places (the order of successes and failures does not matter), and each of the $^nC_k = \binom{n}{k}$ points has probability $p^k q^{n-k}$. Hence the probability of exactly k successes in n trials is

$$b(k; n, p) = \binom{n}{k} p^k q^{n-k} . \qquad (5.24)$$

This function is known as the binomial distribution, because the terms are those of the expansion of $(a + b)^n$ (cf. §4.3).

Bernoulli trials are easily generalized to more than two outcomes. If the probability of realizing an outcome E_i is p_i ($i = 1, 2, ..., r$) subject only to the condition

$$p_1 + p_2 + \cdots + p_r = 1 , \qquad (5.25)$$

then the probability that in n trials E_1 occurs k_1 times, E_2 occurs k_2 times etc. is

$$\frac{n!}{k_1! k_2! \ldots k_r!} p_1^{k_1} p_2^{k_2} \cdots p_r^{k_r} \qquad (5.26)$$

---

[2]Stochastic independence is formally defined via the condition

$$P\{AH\} = P\{A\}P\{H\} , \qquad (5.23)$$

which must hold if the two events A and H are stochastically (sometimes called statistically) independent.

where

$$k_1 + k_2 + \cdots + k_r = n \ . \tag{5.27}$$

The reader can readily verify that a plot of b vs k is a hump whose central term occurs at $m = [(n+1)p]$, where the notation $[x]$ signifies 'the largest integer not exceeding x'.

An important practical case arises where n is large, and p small, such that the product $np = \lambda$ is of moderate size (~ 1). The distribution can then be simplified:

$$b(k; n, p) = \binom{n}{k} \frac{\lambda^k}{n} \left(1 - \frac{\lambda}{n}\right)^{n-k} = \frac{\lambda^k}{k!} \left(1 - \frac{\lambda}{n}\right)^{n-k} \frac{n(n-1)...(n-k+1)}{n^k} .$$

Now $(1 - \lambda/n)^{n-k} \approx e^{-\lambda}$ and $n(n-1)...(n-k+1)/n^k \approx 1$, hence

$$b(k; n, p) \approx \frac{\lambda^k}{k!} e^{-\lambda} = p(k; \lambda) \ , \tag{5.28}$$

which is called the Poisson approximation to the binomial distribution. But if λ is fixed, then $\sum p(k; \lambda) = 1$, hence $p(k; \lambda)$, the probability of exactly k successes occurring, is a distribution in its own right, called the Poisson distribution. It is of great importance in nature, describing processes lacking memory.

The probability $f(k; r, p)$ that exactly k failures precede the rth success (i.e. exactly k failures among $r + k - 1$ trials followed by success) is

$$f(k; r, p) = \binom{r + k - 1}{k} p^r q^k = \binom{-r}{k} p^r (-q)^k, \ k = 0, 1, 2, \ldots \tag{5.29}$$

Iff[3]

$$\sum_{k=0}^{\infty} f(k; r, p) = 1 \ , \tag{5.30}$$

the possibility that an infinite sequence of trials produces fewer than r successes can be discounted since by the binomial theorem

$$\sum_{k=0}^{\infty} \binom{-r}{k} (-q)^k = p^{-r} \tag{5.31}$$

which equals 1 when multiplied by p^r. The sequence $f(k; r, p)$ is called the negative binomial distribution.

---

[3]If and only if.

5.3 Moments of distributions

A *random variable* is 'a function defined on a sample space' (for example, the number of successes in n Bernoulli trials). A unique rule associates a number $\mathbf{X}$ with any sample point. The aggregate of all sample points on which $\mathbf{X}$ assumes the fixed value x_j forms the event that $\mathbf{X} = x_j$, with probability $P\{\mathbf{X} = x_j\}$.[4] The function $f(x_j) = P\{\mathbf{X} = x_j\}$ is called the (probability) distribution of the random variable $\mathbf{X}$.[5] Joint distributions are defined for two or more variables defined on the same sample space. For two variables, $p(x_j, y_k) = P\{\mathbf{X} = x_j, \mathbf{Y} = y_k\}$ is the joint probability distribution of $\mathbf{X}$ and $\mathbf{Y}$.

The mean, average or expected value of $\mathbf{X}$ is defined by

$$\mu = \mathbf{E}(\mathbf{X}) = \sum x_k f(x_k) \tag{5.33}$$

provided that the series converges absolutely. The expectation of the sum (or product) of random variables is the sum (or product) of their expectations. Proofs are left to the reader.

Any function of $\mathbf{X}$ may be substituted for $\mathbf{X}$ in (5.33), with the same proviso of series convergence. The expectations of the rth powers of $\mathbf{X}$ are called the rth moments of $\mathbf{X}$ about the origin.[6] Since $|\mathbf{X}|^{r-1} \leq |\mathbf{X}|^r + 1$, if the rth moment exists, so do all the preceding ones. The expectation of the square of $\mathbf{X}$'s deviation from its mean value has a special name, the variance[7]

$$\sigma^2 = \mathrm{Var}(\mathbf{X}) = \mathbf{E}((\mathbf{X} - \mathbf{E}(\mathbf{X}))^2) = \mathbf{E}(\mathbf{X}^2) - \mathbf{E}(\mathbf{X})^2 . \tag{5.34}$$

Its positive square root σ is called the standard deviation of $\mathbf{X}$, hinting at its use as a rough measure of spread. The mean and variance (i.e. the first and second moments) provide a convenient way to normalize (render dimensionless) a random variable, viz.

$$\mathbf{X}^* = \frac{\mathbf{X} - \mu}{\sigma} . \tag{5.35}$$

---

[4] $\mathbf{X}$ may assume the values $x_1, x_2, \ldots$ (i.e. the range of $\mathbf{X}$).
[5] The distribution function $F(x)$ of $\mathbf{X}$ is defined by

$$F(x) = P\{\mathbf{X} \leq x\} = \sum_{x_j \leq x} f(x_j) \tag{5.32}$$

i.e. a non-decreasing function tending to 1 as $x \to \infty$.
[6] Notice the mechanical analogies: centre of gravity as the mean of a mass and moment of inertia as its variance.
[7] Older literature uses the term 'dispersion'.

Problem. Calculate the means and variances of the binomial and Poisson distributions.

5.3.1 Runs

Studies of the statistical properties of DNA etc. often start by stating the total numbers of the four bases A,C,T,G. This information entirely neglects information on the order in which they occur. The theory of the distribution of runs is one way of handling this information. A run is defined as a succession of similar events preceded and succeeded by different events; the number of elements in a run will be referred to as its length. The number of runs of course equals the number of unlike neighbours.

Here we shall only derive the distribution of runs of two kinds of elements. More complicated results may be found by reference to Mood's 1940 paper.

Let the two kinds of elements be a and b (they could be purines and pyrimidines), and let there be n_1 as and n_2 bs, with $n_1 + n_2 = n$. r_{1i} will denote the number of runs of a of length i, with $\sum_i r_{1i} = r_1$ etc. It follows that $\sum i r_{1i} = n_1$ etc. Given a set of as and bs, the numbers of different arrangements of the runs of a and b are given by multinomial coefficients and the total number of ways of obtaining the set $r_{ji}(j = 1, 2; i = 1, 2, \ldots, n_1)$ is

$$N(r_{ji}) = \left[\begin{array}{c} r_1 \\ r_{1i} \end{array} \right] \left[\begin{array}{c} r_2 \\ r_{2i} \end{array} \right] F(r_1, r_2) \tag{5.36}$$

where the special function $F(r_1, r_2)$ is the number of ways of arranging r_1 objects of one kind and r_2 objects of another so that no two adjacent objects are of the same kind (see Table 5.1).

$\mid r_1 - r_2 \mid$	$F(r_1, r_2)$
> 1	0
1	1
0	2

Table 5.1: Values of the function $F(r_1, r_2)$.

Since there are $\binom{n}{n_1}$ possible arrangements of the as and bs, the distribution of the r_{ji} is

$$P(r_{ji}) = \frac{N(r_{ji})F(r_1, r_2)}{\binom{n}{n_1}}. \tag{5.37}$$

5.3.2 The hypergeometric distribution

Continuing the notation of the previous subsection, consider choosing r elements at random from the binary mixture of as and bs. What is the probability q_k that the group will contain exactly k as? It must necessarily contain $r - k$ bs, and the two types of elements can be chosen in $\binom{n_1}{k}$ and $\binom{n-n_1}{r-k}$ ways respectively. Since any choice of k as can be combined with any choice of $r - k$ bs,

$$q_k = \frac{\binom{n_1}{k}\binom{n-n_1}{r-k}}{\binom{n}{r}} . \tag{5.38}$$

This system of probabilities is called the hypergeometric distribution (because the generating function of q_k is expressible in terms of hypergeometric functions). Many combinatorial problems can be reduced to this form.

Problem. A protein consists of 300 amino acids, of which it is known that there are two cysteines. A 50-mer fragment has been prepared. What are the probabilities that 0, 1 or 2 cysteines are present in the fragment?

5.3.3 Multiplicative processes

Many natural processes are *random additive processes*. For example, a displacement is the sum of random steps (to the left or to the right in the case of the 1-dimensional random walk (cf. Ch. 6). The probability distribution of the net displacement after n steps is the binomial function. The central limit theorem guarantees that this distribution is Gaussian as $n \to \infty$, a universal property of random additive processes.

Although their formalism is less familiar, random multiplicative processes are not less common in nature. An example is rock fragmentation. From an initial value x_0, the size of a rock undergoing fragmentation evolves as $x_0 \to x_1 \to x_2 \to \dots \to x_N$. If the size reduction factor

$$r_n = \frac{x_n}{x_{n-1}} \tag{5.39}$$

is less than one, we have

$$x_N = x_0 \prod_{k=1}^{N} r_k . \tag{5.40}$$

Extreme events, though exponentially rare, are exponentially different. Hence *the average is dominated by rare events*. This is quite different from the more intuitively acceptable random additive processes. If the phenomenon is of that type, the more measurements one can take, the better the estimate of its

value. But if the phenomenon is a random multiplicative process, as one increases the number of measurements, the estimate of the mean will fluctuate more and more, before, ultimately, converging to a stable value. Since multiplication is equivalent to adding logarithms, it comes as no surprise that the distribution is lognormal, i.e. $\ln p = \sum \ln p_i$, and the average value (expectation) of p is

$$\bar{p} = \sum_{n=0}^{N} (Nn) p^n q^{N-n} \ . \tag{5.41}$$

5.4   Likelihood

The search for *regularities* in nature has already been mentioned as the goal of scientific work. Often, these regularities are framed in terms of *hypotheses*.[8] With hypotheses, which may eventually become theories, laws and relations acquire more than immediate validity and relevance.

In observing the natural world, one encounters "deterministic" events, characterized by rather clear relationships between the quantities measured compared with the experimental uncertainties, and more uncertain events with statistical outcomes (such as coin tossing, or Mendelian gene segregation). The latter raise the general problem of how to assess the relative merits of alternative hypotheses in the light of the observed data. Statistics concerns itself with tests of significance, and with estimation, i.e. seeking acceptable values for the parameters of the distributions specified by the hypotheses.

The method of support proposes that

posterior support = prior support + experimental support ,

and

$$\text{information gained} = \log \frac{\text{posterior probability}}{\text{prior probability}} \ .$$

Two rival approaches to estimation have arisen: the theory of inverse probability (due to Laplace), in which the probabilities of causes, i.e. the hypotheses, are deduced from the frequencies of events, and the method of likelihood (due to Fisher). In the theory of inverse probability these probabilities

---

[8]Strictly speaking, one should rather refer to propositions. A hypothesis is an asserted proposition, whereas at the beginning of an investigation it would be better to start with considered propositions, to avoid prematurely asserting what one wishes to find out. Unfortunately the use of the term 'hypothesis' seems to have become so well-established that we may risk confusion if we avoid using the word.

are interpreted as quantitative and absolute measures of belief. Although it still has its adherents, the system of inference based on inverse probability suffers from the weakness of supposing that hypotheses are selected from a continuum of infinitely many hypotheses. The prior probabilities have to be invented, for example by imagining a chance setup, in which case the model is a private one and violates the principle of public demonstrability. Alternatively one can apply Laplace's 'Principle of Insufficient Reason', according to which each hypothesis is given the same probability if there are no grounds to believe otherwise. Conceptually this viewpoint is rather hard to accept. Moreover, if there are infinitely many equiprobable hypotheses, then each one has an infinitesimal probability of being correct.

Bayes' theorem (5.18) may be applied to the weighting of hypotheses if and only if the model adopted includes a chance setup for the generation of hypotheses with specific prior probabilities. Without that, the method becomes one of inverse probability. Equation (5.18) is interpreted as equating the posterior probability of the hypothesis E_k (after having acquired data A) to our prior estimate of the correctness of E_k (i.e. before any data was acquired), $P\{E_k\}$, multiplied by the prior probability of obtaining the data given the hypothesis (i.e. the likelihood, see below), the product being normalized by dividing by the sum over all hypotheses.

A fundamental critique of Bayesian methods is that the Bayes-Laplace approach regards hypotheses as being drawn at random from a population of hypotheses, a certain proportion of which is true. 'Bayesians' regard it as a strength that they can include prior knowledge, or rather prior states of belief, in the estimation of the correctness of a model. Since that appears to introduce a wildly fluctuating subjectivity into the calculations, it seems more reasonable to regard that as a fatal weakness of the method.[9]

To reiterate: our purpose is to find what is the most likely explanation of a set of observations, i.e. a description which is simpler, hence shorter, than the set of facts observed to have occurred.[10]

The three pillars of statistical inference are:

1. A statistical model, i.e. that part of the description that is not at present in question (this corresponds to K in equation 2.12).

2. The data: that which has been observed or measured (unconditional information);

---

[9]As Fisher and others have pointed out, it is not strictly correct to associate Bayes with the inverse probability method. Bayes' doubts as to its validity led him to withhold publication of his work (it was published posthumously).

[10]Sometimes brevity is taken as the main criterion. This is the minimum description length (MDL) approach.

3. The statistical hypothesis: the attribution of particular values to unknown parameters of the model (i.e. those under investigation) (conditional information);

   The preferred values of those parameters are then those that maximize the
likelihood of the model, likelihood being defined in the following paragraph.

Definition.   The likelihood $L(H|R)$ of the hypothesis H given data R and
a specific model is proportional to $P(R|H)$, the constant of proportionality
being arbitrary, but constant in any one application (i.e. with the same model
and same data, but different hypotheses).

   The arbitrariness of the constant of propagation is of no concern, since
in practice likelihood ratios are taken, as in the following definition.

Definition.   The likelihood ratio of two hypotheses on some data is the
ratio of their likelihoods on that data. It will be denoted as $L(H_1, H_2|R)$.
The likelihood ratio of two hypotheses on independent sets of data may be
multiplied together to form the likelihood ratio on the combined data, i.e.

$$L(H_1, H_2|R_1 \& R_2) = L(H_1, H_2|R_1) \times L(H_1, H_2|R_2) \, . \qquad (5.42)$$

   The fundamental difference between probability and likelihood is that
in the inverse probability approach R is variable and H constant, whereas
in likelihood H is variable and R constant. In other words likelihood is
predicated on a fixed R.

   We shall sometimes need to recall that if R_1 and R_2 are two possible,
mutually exclusive, results, and $P\{R|H\}$ is the probability of obtaining the
result R given H, then

$$P\{R_1 \text{ or } R_2|H\} = P\{R_1|H\} + P\{R_2|H\} \qquad (5.43)$$

and

$$P\{R_1 \text{ and } R_2|H\} = P\{R_1|H\}P\{R_2|H\} \, . \qquad (5.44)$$

The method of likelihood reposes on the definitions of likelihood *per se* and
of the likelihood ratio.

Example. The problem is to determine the probability that a baby will be a boy. We take a binomial model (cf. §5.2.3) for the occurrence of boys and girls in a family of two children; we have two sets of data, R_1: one boy and one girl, and R_2: two boys; and we have two hypotheses, H_1: the probability p of a birth being male born equals $\frac{1}{4}$, and H_2: $p = \frac{1}{2}$. Hence

$P\{R\|H\}$	R_1	R_2
H_1	$2p(1-p) = \frac{3}{8}$	$p^2 = \frac{1}{16}$
H_2	$2p(1-p) = \frac{1}{2}$	$p^2 = \frac{1}{4}$

By inspection, $P\{R|H\}$ for H_2 exceeds that for H_1 for both sets of data, from which we may infer that H_2 is better supported by the data.

The concept of likelihood ratio can easily be extended to continuous distributions, i.e. $P\{R|H\}$ becomes a probability density. The likelihood ratio is computed for the distribution with respect to one value chosen arbitrarily and the maximum is sought for. Usually it is better to work in logarithms, and the support $\mathfrak{S}$ is defined as the logarithm of the likelihood, viz.

$$\mathfrak{S}(p) = \log L(p) \tag{5.45}$$

The curvature of $\mathfrak{S}(p)$ at its maximum has been called the information, and its reciprocal is a natural measure of the uncertainty about p, i.e. the width of the peak is inversely related to the degree of certainty of the estimation.

The method of maximum likelihood provides the ability to deliver a conclusion compatible with the given evidence.

5.4.1 The maximum entropy method

Consider the problem of deducing the positions of stars and galaxies from a noisy map of electromagnetic radiation intensity. One should have an estimate for the average noise level: the simple treatment of such a map is to reject every feature greater than the mean noise level, and accept every one that is greater. Such a map is likely to be a considerably distorted version of reality.[11]

The maximum entropy method can be considered as a heuristic drill for applying D. Bernoulli's maxim: "Of all the innumerable ways of dealing with errors of observation, one should choose the one which has the highest degree of probability for the complex of observations as a whole". In effect, it is a generalization of the method of maximum likelihood.

---

[11]Implicitly, Platonic reality is meant here.

Firstly, the experimental map must be digitized both spatially and with respect to intensity: i.e. it is encoded as a finite set of pixels, each of which may assume one of a finite number of density levels. Let that density be m_j at the jth pixel. Then, random maps are generated and compared with the data. All those inconsistent with the data (with due regard to the observational errors) are rejected. The commonest map remaining is the most likely representation. This process is the constrained maximization of the configurational entropy $-\sum m_j \log m_j$ (the unconstrained maximization would simply lead to a uniform distribution of density over the pixels). Maximum entropy image restoration yields maximum information in Shannon's sense.

Chapter 6

Randomness and complexity

Randomness is a concept which is deeply entangled with bioinformatics. A random sequence cannot convey information, in the sense that it could be generated by a recipient merely by tossing a coin. Randomness is therefore a kind of 'null hypothesis': a random sequence of symbols is a sequence lacking all contraints limiting the variety of choice of successive symbols selected from a pool with constant composition (i.e. an ergodic source). Such a sequence has maximum entropy in the Shannon sense, i.e. it has minimum redundancy.

If we are using such an ideally random sequence as a starting point for assessing departures from randomness, it is important to be able to recognize this ideal randomness. How easy is this task? Consider the following three sequences:

11111111111111111111111111111111
01010101010101010101010101010101
10010100010100101010111101001010101010

each of which could have been generated by tossing a coin. According to the results from the previous two chapters, all three outcomes, indeed any sequence of 32 ones and zeros, have equal probability of occurrence, namely $1/2^{32}$. Why do the first two not 'look' random? Kolmogorov supposed that the answer might belong to psychology; Borel even asserted that the human mind is unable to simulate randomness (presumably the ability to recognize patterns was—and is—important for our survival). Yet apparent pattern is also present in random sequences: van der Waerden has proved that in every infinite binary sequence at least one of the two symbols must occur in arithmetical progressions of every length. Hence the first of the above three sequences would be an unexceptionable occurrence in a much longer random sequence. In fact, whether a given sequence is random is formally undecidable. At best, then, we can hope for heuristic clues to the possible absence of randomness, and hence presumably meaning, in a gene sequence.

In anticipation of the following sections, we can already note that incompressibility, i.e. the total absence of regularities, forms a criterion of randomness. This criterion uses the notion of algorithmic complexity. The first sequence can be generated by the brief instruction "write '1' 32 times", and the second by the only marginally longer statement "write '01' 16 times", whereas the third, which was generated by blindly tapping on a keyboard, has no apparent regularity.

'Absence of pattern' corresponds to the dictionary synonym 'haphazard' (cf. the French expression 'au hasard'). By counting the number of ones and zeros in a long segment of the third sequence, we can obtain an estimate of the probability of occurrence of each symbol. 'Haphazard' then means that the choice of each successive symbol is made independently, without reference to the preceding symbol or symbols, in sharp contrast to the second sequence, which could also be generated by the algorithm 'if the preceding symbol is 1, write 0, otherwise write 1' operating on a starting seed of 1 or 6.

Note how closely this exercise of algorithmic compression is related to the general aim of science: to find the simplest set of axioms that will enable all the observable phenomena studied by the branch of science concerned to be explained (an empirical fact being 'explained' if the propositions expressing it can be shown to be a consequence of the axioms constituting the scientific theory underpinning that branch). For example, Maxwell's equations turned out to be suitable for explaining the phenomena of electromagnetism.[1]

The meaning of randomness as denoting independence from what has gone before is well captured in the familiar expression 'random access memory', the significance being that a memory location can be selected arbitrarily (cf. the German 'beliebig', at whim), as opposed to a sequential access memory, whose elements can only be accessed one after the other. Mention of memory brings to mind the fact that successive independent choices implies the absence of memory in the process generating those choices.

The validity of the above is independent of the actual probabilities of

---

[1]An obvious corollary of this association of randomness with algorithmic compressibility is that there is an intrinsic absurdity in the notion of an algorithm for generating random numbers, such as those included with many compilers and other software packages. These computer-generated pseudorandom numbers generally pass the usual statistical tests for randomness, but little is known about how their nonrandomness affects results obtained using them. Quite possibly the best heuristic sources of (pseudo)random digits are the successive digits of irrational numbers like π or $\sqrt{2}$. These can be generated by a deterministic algorithm and of course are always the same, but in the sense that one cannot jump to (say) the hundredth digit without computing those preceding it, they do fulfil the criteria of haphazardness.

choosing symbols, i.e. they may be equal or unequal. Although in many organisms it turns out that the frequencies of occurrence of all four bases are in fact equal (to the extent that we can make such an assertion for any finite sequence), this is by no means universal, it being well known that thermophilic bacteria have more C...G base pairs than A..T in their genes, since the former, being linked by three hydrogen bonds, are more stable than the latter, which only have two (cf. figure 10.3). Yet we can still speak of randomness in this case. In binary terms, it corresponds to unequal probabilities of heads or tails, and the sequence may still be algorithmically incompressible, i.e. it cannot be generated by any means shorter than the process actually used to generate it.

We have previously stated that bioinformatics could be considered to be the study of the departures from randomness of DNA. We are shown a sequence of DNA: is it random? We want to be able to quantify its departure from randomness. Presumably those sequences belonging to viable organisms, or even to their individual proteins or promoter sequences, are not random. What about introns, and inter-genome sequences? If they are indeed "junk", as is sometimes (facetiously?) asserted, then we might well expect them to be random. Even if they started their existence as nonrandom sequences, they may have been randomized since they would be subject to virtually no selection pressure. Mutations are supposed to be random and occur at random places. The opposite procedure would be that all DNA sequences started as random ones, and then natural selection eliminated many according to some systematic criterion, and therefore the extant collection of the DNA of viable organisms on this planet is not random. Can we, then, say anything about the randomness or otherwise of an individual sequence taken in isolation?

Similar considerations apply to proteins. Given a collection of amino acid sequences of proteins (which, to be meaningful, should come from the same genome) we can assess the likelihood that they arose by chance, and the degree of their departures from randomness.

All such sequences can be idealized as sequences of Bernoulli trials (see §5.2.3), which are themselves abstractions of a coin tossing experiment. Since order does not matter in determining the probability of a given overall outcome, fifty heads followed by fifty tails has the same probability of occurring as fifty alternations of heads and tails, which again is no less probable than a particular realization in which the heads and tails are "randomly" mixed.

Any non-binary sequence can of course be encoded in binary form. Typical procedures for biological sequences (amino acids or nucleotides) are to consider nucleotides as purines (0) or pyrimidines (1), or amino acids as hydrophobic (apolar) or hydrophilic (polar) residues, etc. (cf. Markov's en-

coding of poetry as a sequence of vowels and consonants). Alternatively, the
nucleotides could constitute a sequence in base 4 (A ≡ 0, C ≡ 1, T ≡ 2, G
≡ 3), which can then be converted to base 2.

It is a commonly held belief that after a long sequence of heads (say), the
opposite result (tails) becomes more probable. There is no empirical support
for this assertion in the case of coin tossing. In other situations in which
the outcome depends on selecting elements from a finite reservoir, however,
clearly this result must hold. Thus, if a piece of DNA is being assembled
from a soup of base monomers at initially equal concentrations, if by chance
the sequence starts out by being poor in A, say, then later on this must
be compensated by enrichment, (chain elongation ends when all available
nucleotides have been consumed).

Formal notions of randomness

In order to proceed further, we need to more carefully understand what we
mean by randomness. Despite the fact that the man in the street supposes
that he has a good idea of what it means, randomness is a rather delicate
concept. The toss of an unbiased coin is said to be random: the probability
of heads or tails is 0.5. We cannot assess the randomness of a single result,
but we can assess the probability that a sequence of tosses is random. So
maybe we can anwer the question whether a given individual sequence is
random. The three main notions of randomness are:[2]

1. Stochasticity, or frequency stability, associated with von Mises, Wald,
   and Church;[3]

2. Incompressibility or chaoticity, associated with Solomonoff, Kolmogorov,
   and Chaitin;[4] and

---

[2]After Volchan.

[3]Von Mises called the random sequences in accord with this notion 'collectives'. It was
subsequently shown that the collectives were not random enough (see Volchan for more
details). For example, the number $0.0123456789101112131415161718192021\ldots$ satisfied
von Mises' criteria, but is clearly computable.

[4] The Kolmogorov-Chaitin definition of the descriptive or algorithmic complexity $K(s)$
of a symbolic sequence s with respect to a machine M is given by

$$K(s) = \begin{cases} \infty & \text{if there is no } p \text{ such that } M(p) = s , \\ \min\{|p| : M(p) = s\} & \text{otherwise .} \end{cases} \tag{6.1}$$

This means that $K(s)$ is the size of the smallest input program p that prints s and then
stops when input into M. In other words, it is the length of the shortest (binary) program
that describes (codifies) s. Insofar as M is usually taken to be a universal Turing machine,
the definition is machine-independent.

3. Typicality, associated with Martin-Löf (and essentially coincident with incompressibility).

6.1 Random processes

A process characterized by a succession of values of a characteristic parameter y is called random if y does not depend in a completely definite way on the independent variable, usually (laboratory) time t, but in the context of sequences the independent variable could be the position along the sequence. A random process is therefore essentially different from a causal process (cf. §5.1). It can be completely defined by the set of probability distributions $W_1(yt)dy$,the probability of finding y in the range $(y, y+dy)$ at time t, $W_2(y_1t_1, y_2t_2)dy_1dy_2$,the joint probability of finding y in the range (y_1, y_1+dy_1) at time t_1 and in the range (y_2, y_2+dy_2) at time t_2, and so on for triplets, quadruplets of values of y.

If there is an unchanging underlying mechanism, the probabilities are stationary and the distributions can be simplified as $W_1(y)dy$, the probability of finding y in the range $(y, y+dy)$; $W_2(y_1y_2t)dy_1dy_2$, the joint probability of finding y in the ranges (y_1, y_1+dy_1) and (y_2, y_2+dy_2) separated by an interval $t = t_2-t_1$; etc. Experimentally a single long record $y(t)$ can be cut into pieces (which should be longer than the longest period supposed to exist), rather than carrying out measurements on many similarly prepared systems. This equivalence of time and ensemble averages is called ergodicity. Note, however, that many biological systems appear to be frozen in small regions of state space, as a glass, and hence are non-ergodic (cf. §3.4).

Notice some of the difficulties inherent in the above description. For example, we referred to 'an unchanging underlying mechanism', yet at the same time asserted that a random process is one which does not depend in a completely definite way on the independent variable. Yet who would deny that the coin, whose tossing generates that most archetypical of random sequences, does not follow Newton's laws of motion? This apparent paradox is shown to be a consequence of dynamic chaos (§7.3).

If successive values of y are not correlated at all, i.e.

$$W_2(y_1t_1, y_2t_2) = W_1(y_1t_1)W_1(y_2t_2) \tag{6.2}$$

etc., all information about the process is completely contained in W_1, and the process is called a purely random process.

6.2   Markov chains

In the previous section we considered 'purely random' processes in which successive values of a variable, y, are not correlated at all. If, however, the next step of a process depends on its current state, i.e.

$$W_2(y_1 y_2 t) = W_1(y_1) P_2(y_2 | y_1 t) \tag{6.3}$$

where P_2 denotes the conditional probability (that y is in the range (y_2, y_2+dy_2) after having been at y_1 at a time t earlier), we have a Markov chain.

Definition. A sequence of trials with possible outcomes **a** (possible states of the system), an initial probability distribution $\mathbf{a}^{(0)}$, and (stationary) transition probabilities defined by a stochastic matrix P is called a Markov chain.[5]
    The probability distribution for an r-step process is

$$\mathbf{a}^{(r)} = \mathbf{a}^{(0)} P^r . \tag{6.4}$$

If the first m steps of a Markov process lead from a_j to some intermediate state a_i, then the probability of the subsequent passage from a_i to a_k does not depend on the manner in which a_i was reached, i.e.

$$p_{jk}^{(m+n)} = \sum_i p_{ji}^{(m)} p_{ik}^{(n)} \tag{6.5}$$

where $p_{jk}^{(n)}$ is the probability of a transition from a_j to a_k in exactly n steps (this is a special case of the Chapman-Kolmogorov identity).
    If upon repeated application of P the distribution **a** tends to an unchanging limit (i.e. an equilibrium set of states) that does not depend on the initial state, the Markov chain is said to be ergodic, and we can write

$$\lim_{r \to \infty} P^r = Q \tag{6.6}$$

where Q is a matrix with identical columns.[6] Now

$$PP^n = P^n P = P^{n+1} \tag{6.7}$$

and if Q exists it follows, by letting $n \to \infty$, that

$$PQ = QP = Q \tag{6.8}$$

---

[5]In this book (and others), the columns of stochastic matrices sum to unity; in the rest of the literature one finds stochastic matrices arranged such that the rows sum to unity.
[6]As for the transition matrix for a zeroth order chain, i.e. independent trials.

from which Q (giving the stationary probabilities, i.e. the equilibrial distribution of **a**) can be found.

If all the transitions of a Markov chain are equally probable, then there is a complete absence of constraint; the process is purely random (a zeroth order chain). Higher order Markov processes have already been dealt with (§2.2).

A Markov chain represents an automaton (cf. §7.1.1) working incessantly. If the transformations were determinate (i.e. all entries in the transition matrix were zero or one) then the automaton would reach an attractor after a finite number of steps. The non-determinate transformation can, however, continue indefinitely although if any diagonal element is unity it will get stuck there). If chains are nested inside one another, one obtains a hidden Markov model (HMM, see §12.5.3): suppose that the transformations accomplished by an automaton are controlled by a parameter that can take values a_1 or a_2, say. If a_1 is input, the automaton follows one matrix of transitions, if a_2, another set. The HMM is created if transitions between a_1 and a_2 are also Markovian.

Markov chain Monte Carlo (MCMC) is used when the number of unknowns is itself an unknown.

Problem. Prove that if P is stochastic, then any power of P is also stochastic.

6.2.1 The entropy of a Markov process

The entropy of a Markov process is the 'average of an average' in the following sense: the entropy of the transitions (i.e. the weighted variety of the transitions) can be found from each column of the stochastic matrix according to equation (2.5). The (informational) entropy of the process as a whole is then the weighted average of these entropies, the weighting being given by the equilibrium distribution of the states.

Problem. Consider the three state Markov chain

↓	1	2	3
1	0.1	0.5	0.3
2	0.9	0.0	0.3
3	0.0	0.5	0.4

and calculate (i) the equilibrium proportions of the states 1, 2 and 3, and (ii) the average entropy of the entire process.

6.3   Random walks

Consider an agent on a line susceptible to step right with probability p and left with probability $q = 1 - p$. We can encode the walk by writing $+1$ for a right step and -1 for a left step. Many processes can be mapped onto the random walk, for example a nucleic acid sequence, with purines $\equiv -1$ and pyrimidines $\equiv +1$. If the walk is drawn in Cartesian coordinates as a polygon with the number of steps ('time') along the horizontal axis and the displacement along the vertical axis, then if s_k is the partial sum of the first k steps,

$$s_k - s_{k-1} = \pm 1, \qquad s_0 = 0, \qquad s_n = n(p - q), \qquad (6.9)$$

where n is the length of the path.

Definition.   Let $n > 0$ and x be integers. A path $(s_1, s_2, \ldots, s_n)$ from the origin to the point (n, x) is a polygonal line whose vertices have abscissae $0, 1, \ldots, n$ and ordinates $s_0, s_1, \ldots, s_n$ satisfying $s_k - s_{k-1} = \epsilon_k = \pm 1$, $s_0 = 0$, $s_n = p - q$ (where p and q are now the numbers of symbols, $p + q = n$), with $s_n = x$.

There are 2^n paths of length n, but a path from the origin to an arbitrary point (n, x) exists only if n and x satisfy

$$n = n(p + q), \qquad x = n(p - q). \qquad (6.10)$$

In this case the np positive steps can be chosen from among the n available places in

$$N_{n,x} = \binom{p + q}{p} = \binom{p + q}{q} \qquad (6.11)$$

ways. The average distance travelled after n steps is $\sim n^{1/2}$. The random walk is of course an example of a Markov chain.

Problem.   Write out the transition matrix for a random walk in one dimension.

6.4   Noise

It might be thought that 'noise' is the ultimate random, uncorrelated process. In reality, however, noise can come in various 'colours' according to the exponent of its power spectrum.

Let $x(t)$ describe a fluctuating quantity. It can be characterized by the two-point autocorrelation function

$$C_x(n) = \sum_{j=1}^{N} x_j x_{j-n} \qquad (6.12)$$

(in discrete form), where n is the position along a nucleic acid or protein sequence of N elements, and the spectrum or amplitude spectral density

$$A_x(m) = \sum_{j=-\infty}^{\infty} x_j e^{-2\pi i m} , \qquad (6.13)$$

whose square is the power spectrum or power spectral density:

$$S_x(m) = |A_x(m)|^2 , \qquad (6.14)$$

where m is sequential frequency. The autocorrelation function and the power spectrum are just each other's Fourier transforms (the Wiener-Kintchin relations, applicable to stationary random processes).

A truly random process ('white noise', $w(t)$) should have no correlations in time. Hence

$$C_w(\tau) \propto \delta(\tau) \qquad (6.15)$$

and

$$S_w(f) \propto 1 ; \qquad (6.16)$$

the power spectrum is convergent at low frequencies, but if one integrates up from some finite frequency towards infinity, one finds a divergence: there is an infinite amount of power at the highest frequencies, i.e. a plot of $w(t)$ is infinitely choppy, and the instantaneous value of $w(t)$ is undefined!

White noise is also called Johnson (who first measured it experimentally, in 1928) or Nyquist (who first derived its power spectrum theoretically) noise. It is characteristic of the voltage across a resistor measured at open circuit and is due to the random motions of the electrons. The integral of white noise,

$$B(t) = \int w(t) dt , \qquad (6.17)$$

corresponds to a random walk or Brownian motion (hence 'brown noise'). Its power spectrum is

$$S_B(f) \propto 1/f^2 , \qquad (6.18)$$

i.e. it is convergent when integrating to infinity, but divergent when integrating down to zero frequency. In other words, the function has a well-defined value at each point, but wanders ever further from its initial value at longer and longer times, i.e. it does not have a well-defined mean value.

If current is flowing across a resistor, then the power spectrum of the voltage fluctuations $S_F(f) \propto 1/f$ ('$1/f$ noise', sometimes called "fractional gaussian noise" (FGN)), as a special case of fractionally integrated white noise. FGNs are characterized by a parameter F, the mean distance travelled in the process described by its integral $G_F(t) = \int x_F(t)dt$ is proportional to t^F, and the power spectrum $S_G(f) \propto 1/f^{2F-1}$. White noise has $F = \frac{1}{2}$, and $1/f$ noise has $F = 1$. It is divergent when integrated to infinite frequency, and when integrated to zero frequency, but the divergences are only logarithmic. $1/f$ noise exhibits very long range correlations, for which the physical reason is still a mystery. Many natural processes exhibit $1/f$ noise.

6.5 Complexity

The notion of complexity occurs rather frequently in biology, where one often refers to the complexity of this or that organism. Several procedures for ascribing a numerical value to it have been devised, but for all that it remains somewhat elusive. When we assert that a mouse is more complex than a bacterium (or than a fly), what do we actually mean? Intuitively the assertion is unexceptionable—most people would presumably readily agree that man is the most complex organism of all. Is our genome the biggest (as may once have been believed)? No. Do we have more cell types than other organisms? Yes, and the mouse has more than the fly, but then complexity becomes merely a synomym for variety. Or does it reflect what we can do? Man alone can create poems, theories, musical compositions, paintings etc. But while one could perhaps compare the complexity of different human beings on that basis, it would be useless for the rest of the living world. Is complexity good or bad? A complex theory that noone apart from its inventor can understand might be impressive, but not very useful. On the other hand we have the notion, again rather intuitive, that a complex organism is more adaptable than a simple one, because it has more possibilities for action, hence it can better survive in a changing environment.[7]

Other pertinent questions are whether complexity is an absolute attribute of an object, or does it depend on the level of detail with which one describes it (in other words, how its description is encoded—an important consideration if one is going to extract a number to quantify complexity)? Every writer

---

[7]If this is so, it then seems rather strange that so much ingenuity is expended by presumably complex people to make their environments more uniform and unchanging, in which case they will tend to lose their competitive advantage.

on the subject seems to introduce his own particular measure of complexity, with a corresponding special name—what do these different measures have in common? Do printed copies of a Shakespeare play have the same complexity as the original manuscript? Does the fiftieth edition have less complexity than the first?

The antonym of complexity is simplicity; the antonym of randomness is regularity. A highly regular pattern is also simple. Does this, then, suggest that complexity is a synonym for randomness?

An important advance was Kolmogorov's notion of algorithmic complexity (also called algorithmic information content, AIC) as a criterion for randomness. As we have seen near the beginning of this chapter, the AIC, $K(s)$, of a string s is the length of the smallest program (running on a universal computing machine) able to print out s. Henceforth we shall mainly consider the complexity of strings: objects can of course be encoded as strings. If there are no regularities, $K(s)$ will have its maximum possible value, which will be roughly equal to the length of the string: no compression is possible; the string has to be printed out verbatim.[8] Hence

$$K_{\max} = |s| . \tag{6.19}$$

Any regularities, i.e. constraints in the choice of successive symbols, will diminish the value of K. We call $K_{\max}$ the unconditional complexity; it is actually a measure of regularity.

This definition leads to the intuitively unsatisfying consequence that the highest possible complexity, the greatest potential information gain, etc. is possessed by a purely random process, which then implies that the output of the proverbial team of monkeys tapping on keyboards is more complex than a Shakespeare play, etc. (the difference would, however, vanish if the letters of the two texts were encoded in such a way that only one symbol was used to encode each letter). What we would like is some quantity that is small for highly regular structures (low disorder), and then increases to a maximum as the system becomes more disordered, and finally falls back to a low value as the disorder approaches pure randomness.

It will be useful to introduce some additional quantities, such as the joint algorithmic complexity $K(s,t)$, the length of the smallest program required to print out the two strings s and t:

$$K(s,t) \approx K(t,s) \lesssim K(s) + K(t) ; \tag{6.20}$$

---

[8]Many considerations of complexity may be reduced to the problem of printing out a number. Thus, the complexity of a protein structure is related to the number specifying the positions of the atoms, or dihedral angles of the peptide groups, which is equivalent to selecting one from a list of all possible conformations; the difficulty of doing that is roughly the same as that of printing out the largest number in that list.

the mutual algorithmic information

$$K(s:t) = K(s) + K(t) - K(s,t) \tag{6.21}$$

(which reflects the ability of a string to share information with another string); conditional algorithmic information (or conditional complexity)

$$K(s|t) = K(s,t) - K(t) \tag{6.22}$$

i.e. the length of the smallest program that can compute s from t; and algorithmic information distance

$$D(s,t) = K(s,t) + K(t|s) \tag{6.23}$$

(the reader may verify that this measure fulfils the usual requirements for a distance).

Adami and Cerf have emphasized that randomness and complexity only exist with respect to a specific defined environment e. Consider the conditional complexity $K(s|e)$. The smallest program for computing s from e will only contain elements entirely unrelated to e, since if they were related, they could be obtained (i.e. deduced) from e with a program tending to size zero. Hence $K(s|e)$ quantifies those elements in s that are random (with respect to e) (if there is no environment, then all strings have the maximum complexity, $K_{\max}$). We can now use the the mutual algorithmic information defined above (equation 6.21) to determine

$$K(s:e) = K_{\max} - K(s|e) \qquad , \tag{6.24}$$

which represents the number of meaningful elements in string s. Nevertheless it might not be practically possible to compute $K(s|e)$ unless one is aware of the coding scheme whereby some of e is encapsulated in s. A possible way to overcome this difficulty is opened where multiple copies of a sequence that have adapted independently to e exist. It may then reasonable be assumed that the coding elements are conserved (and have a nonuniform probability distribution), whereas the noncoding bits are fugitive (and have a uniform probability distribution). The information about e contained in the ensemble S of copies is then $I(S)$ (the Shannon index) $-I(S|e)$. In finite ensembles, the quantity

$$I(S|e) = - \sum_s p(s|e) \log p(s|e) \tag{6.25}$$

can be estimated by sampling the distribution $p(s|e)$.

Computational complexity reflects how the number of elementary operations required to compute a number increases with the size of that number.

Hence, the computational complexity of '011011011011011011...' is of order unity, since one merely has to specify the number of repetitions.

Algorithmic and computational complexity are combined in the concept of logical depth,[9] defined as the number of elementary operations (machine cycles) required to calculate a string from the shortest possible program. Hence the number π, whose specification requires only a short program, has considerable logical depth because that program has to execute many operations to yield π.

Gell-Mann has proposed effective complexity to be proportional to the length of a concise description of a set of an object's regularities, which amounts to the algorithmic complexity of the description of the set of regularities. This prescription certainly fulfils the criterion of correspondence with the intuitive notion of complexity: both a string consisting of one type of symbol and the monkey-text would have no variety in their regularity and hence minimal complexity. One way of assessing the regularities is to divide the object into parts and examine the mutual algorithmic complexity between the parts. The effective complexity is then proportional to the length of the description of those regularities.

A more physically oriented approach has been proposed by Lloyd and Pagels. Their notion of (thermodynamic) depth attempts to measure the process whereby an object is constructed. A complex object is one that is difficult to put together;[10] the average complexity of a state is the Shannon entropy of the set of trajectories leading to that state $(-\sum p_i \log p_i$, where p_i is the probability that the system has arrived at that state by the ith trajectory) and the depth $\mathcal{D}$ of a system in a macroscopic state d is

$$\mathcal{D}(d) = -k \log p_i \ . \tag{6.26}$$

An advantage of this process-oriented formulation is the way in which the complexity of copies of an object can be dealt with: the depth of a copy, or any number of copies, is proportional to the depth of making the original object plus the depth of the copying process.

Problem. A deep notion is generally held to be more meaningful than a shallow one. Could one, then, identify complexity with meaning? Discuss the use of the ways of quantifying complexity, especially effective complexity, as a measure of meaning (cf. §2.3.2).

---

[9]Due to C.H. Bennett.

[10]Cf. the nursery rhyme *Humpty Dumpty sat on a wall / Humpty Dumpty had a great fall / And all the king's horses and all the king's men / Couldn't put Humpty together again.* It follows that Humpty Dumpty had great depth, hence complexity.

Chapter 7

Systems, networks and circuits

Just as we are often interested in events that are composed of many elementary (simple) events, in biology the objects under scrutiny are vastly complex objects composed of many individual molecules (the molecule is probably the most appropriate level of course graining for the systems we are dealing with). Since these components are connected together, they constitute a system. The essence of a system is that it cannot be usefully decomposed into its constituent parts. More formally, following R.L. Ackoff we can assert that two or more objects (which may be entities or activities etc.) constitute a *system* if the following four conditions are satisfied:

1. One can talk meaningfully of the behaviour of the whole of which they are the only parts;

2. The behaviour of each part can affect the behaviour of the whole;

3. The way each part behaves and the way its behaviour affects the whole depends on the behaviour of at least one other part;

4. No matter how one subgroups the parts, the behaviour of each subgroup will affect the whole and depends on the behaviour of at least one other subgroup.

There are various corollaries, one of the most important and practical of which is that a system cannot be investigated by looking at its components individually, or by varying one parameter at a time, as Fisher seems to have been the first to realize. Thus, a *modus operandi* of the experimental scientist inculcated at an early age and reinforced by the laboratory investigation of 'simple systems'[1] turns out to be inappropriate and misleading when applied

---

[1]Here we plead against the use of the terms 'simple system' and 'complex system': the criteria given above imply that no system is simple, and that every system is complex.

to most phenomena involving the living world.

Another corollary is that the concept of feedback, which is usually clear enough to apply to two-component systems, is practically useless in more complex systems.[2]

In this chapter, we shall first consider the approach of general systems theory, largely pioneered by Bertalanffy. This allows some insight into the behaviour of very simple systems with not more than two components, but thereafter statistical approaches have to be used.[3] This is successful for very large systems, in which statistical regularities can be perceived; the most difficult cases are those of intermediate size.

Some properties of networks *per se* will then be examined, and finally the insight of the electrical or electronic engineer will be brought to bear, looking at systems as circuits.

Problem.   Consider various familiar objects, and ascertain using the above criteria whether they are systems.

7.1   General systems theory

Consider a system containing n interacting elements $G_1, G_2, \ldots, G_n$. Let the values of these elements be $g_1, g_2, \ldots, g_n$. For example, if the G denote species of animals, then g_1 could be the number of individual animals of species G_1. The temporal evolution of the system is then described by

$$
\begin{aligned}
\frac{dg_1}{dt} &= \mathcal{G}_1(g_1, g_2, \ldots, g_n) \\
\frac{dg_2}{dt} &= \mathcal{G}_2(g_1, g_2, \ldots, g_n) \\
&\;\;\vdots \\
\frac{dg_n}{dt} &= \mathcal{G}_n(g_1, g_2, \ldots, g_n)
\end{aligned}
\tag{7.1}
$$

where the functions $\mathcal{G}$ include terms proportional to $g_1, g_1^2, g_1^3, \ldots, g_1 g_2, g_1 g_2 g_3, \ldots$ etc. In practice, many of the coefficients of these terms will be close or equal to zero.

---

[2]Even in two component systems its nature can be elusive. For example, as Ashby has pointed out, are we to speak of feedback between the position and momentum of a pendulum? Their interrelation certainly fulfils all the formal criteria for the existence of feedback.

[3]Robinson has recently proved that all possible chaotic dynamics can be approximated in three dimensions.

If we only consider one variable,

$$\frac{dg_1}{dt} = \mathcal{G}_1(g_1) .$$

(7.2)

Expanding gives

$$\frac{dg_1}{dt} = rg_1 - \frac{r}{K}g_1^2 + \cdots$$

(7.3)

where $r > 0$ and $K > 0$ are constants. Retaining terms up to g_1 gives simple exponential growth,

$$g_1(t) = g_1(0)e^{rt}$$

(7.4)

where $g_1(0)$ is the quantity of g_1 at $t = 0$. Retaining terms up to g_1^2 gives

$$g_1(t) = \frac{K}{1 + e^{-r(t-m)}} ,$$

(7.5)

the so-called logistic equation, which is sigmoidal with a unique point of inflexion at $t = m, g_1 = K/2$ at which the tangent to the curve is r, and asymptotes $g_1 = 0$ and $g_1 = K$. r is called the growth rate and K is called the carrying capacity in ecology.

Consider now two objects,

$$\left. \begin{array}{l} dg_1/dt = a_{11}g_1 + a_{12}g_2 + a_{111}g_1^2 + \cdots \\ dg_2/dt = a_{21}g_1 + a_{22}g_2 + a_{211}g_1^2 + \cdots \end{array} \right\}$$

(7.6)

in which the functions $\mathcal{G}$ are now given explicitly in terms of their coefficients a (a_{11}, for example, gives the time in which an isolated G_1 returns to equilibrium after a perturbation). The solution is

$$\left. \begin{array}{l} g_1(t) = g_1^* - h_{11}e^{\lambda_1 t} - h_{12}e^{\lambda_2 t} - h_{111}e^{2\lambda_1 t} - \cdots \\ g_2(t) = g_2^* - h_{21}e^{\lambda_1 t} - h_{22}e^{\lambda_2 t} - h_{211}e^{2\lambda_1 t} - \cdots \end{array} \right\}$$

(7.7)

where the starred quantities are the stationary values, obtained by setting $dg_1/dt = dg_2/dt = 0$, and the λs are the roots of the characteristic equation, which is (ignoring all but the first two terms of the right hand side of equation 7.6)

$$\begin{vmatrix} a_{11} - \lambda & a_{12} \\ a_{21} & a_{11} - \lambda \end{vmatrix} = 0 .$$

(7.8)

Depending on the values of the a coefficients, the phase diagram (i.e. a plot of g_1 vs g_2) will tend to a point (all λ are negative), or a limit cycle (the λ are imaginary, hence there are periodic terms), or there is no stationary state

(λ are positive). Regarding the last case, it should be noted that however large the system, a single positive λ will make one of the terms in (7.7) grow exponentially and hence rapidly dominate all the other terms.

Although this approach can readily be generalized to any number of variables, the equations can no longer be solved analytically and indeed the difficulties become forbidding. Hence one must turn to statistical properties of the system. Equation (7.6) can be written compactly as

$$\dot{\mathbf{g}} = A\mathbf{g} \qquad (7.9)$$

where $\mathbf{g}$ is the vector $(g_1, g_2, \ldots)$, $\dot{\mathbf{g}}$ its time differential, and A the matrix of the coefficients a_{11}, a_{12} etc. connecting the elements of the vector. The binary connectivity C_2 of A is defined as the proportion of nonzero coefficients.[4] In order to decide whether the system is stable or unstable, we merely need to ascertain that none of the roots of the characteristic equation are positive, for which the Routh-Hurwitz criterion can be used without actually having to solve the equation. Gardner and Ashby determined the dependence of the probability of stability on C_2 by distributing nonzero coefficients at random in the matrix A for various values of the number of variables n. They found a sharp transition between stability and instability: for $C < 0.13$, a system will almost certainly be stable, and for $C > 0.13$, almost certainly unstable. For very small n the transition was rather gradual, viz. for $n = 7$ the probability of stability is 0.5 at $C_2 \approx 0.3$, and for $n = 7$, at $C_2 \approx 0.7$.

7.1.1   Automata

We can generalize the Markov chains from §6.2 by writing equation 7.9 in discrete form:

$$\mathbf{g}' = A\mathbf{g} \qquad (7.10)$$

i.e. the transformation A is applied at discrete intervals and $\mathbf{g}'$ denotes the values of g at the epoch following the starting one. The value of g_i now depends not only on its previous value, but also on the previous values of some or all of the other $n - 1$ components. Generalizations to the higher-order coefficients are obvious but difficult to write down; we should recall that application of this approach to the living cell is likely to require perhaps third or fourth order coefficients, but that the corresponding matrices will be extremely sparse.

The analysis of such systems usually proceeds by restricting the values of the g to integers, and preferably to just zero or one (Boolean automata).

---

[4]The ternary connectivity takes into account connexions between three elements, i.e. contains coefficients like a_{123}, etc.

Consider an automaton with just three components, each of which has an output connected to the other two. Equation (7.10) becomes

$$
\begin{pmatrix} g_1 \\ g_2 \\ g_3 \end{pmatrix}' = \begin{pmatrix} 0 & 1 & 1 \\ 1 & 0 & 1 \\ 0 & 0 & 1 \end{pmatrix} \diamondsuit \begin{pmatrix} g_1 \\ g_2 \\ g_3 \end{pmatrix} \tag{7.11}
$$

where $\diamondsuit$ denotes that the additions in the matrix multiplication are to be carried out using Boolean AND logic, i.e. according to table 7.1. Enumerating

input	output
0	0
1	0
2	1

Table 7.1: Truth table for an AND gate.

all possible starting values leads to the following state structure (figure 7.1). The problem at the end of this subsection will help the reader to be convinced that state structure is not closely related to physical structure (the pattern of interconnexions). In fact, to study a system one needs to determine the state structure and know both the interconnexions and the functions of the individual objects (cells).

Most of the work on the evolution of automata (state structures) considers the actual structure (interconnexions) and the individual cell functions to be immutable. For biological systems, this appears to be an oversimplification. Relative to the considerable literature on the properties of various kinds of networks, very little has been done on evolving networks, however.[5]

Problem. Determine the state structure of an automaton if (i) the functions of the individual cells are changed from those represented by (7.11) such that G_1 becomes 1 whenever G_2 is 1, G_2 becomes 1 whenever G_3 is 1, and G_3 becomes 1 whenever G_1 and G_2 have the same value; (ii) keep these functions, but connect G_1's output to itself and G_3, G_2's output to itself, G_1 and G_3, and G_3's output to G_2; and (iii) keep these new interconnexions, but restore the functions to those represented by (7.11) and table 7.1. Compare the results with each other and with figure 7.1.

---

[5] An exception is Érdi and Barna's work on a model of neuron interconnexions, simulating Hebb's rule (traffic on a synapse strengthens it, i.e. increases its capacity).

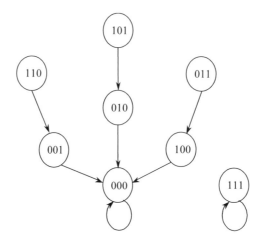

Figure 7.1: State structure of the automaton represented by equation (7.10) and table 7.1.

7.1.2   Cellular automata

This term is usually applied to cells arranged in spatial proximity to each other. The most widely studied ones are only connected to their nearest neighbours. Despite this simplicity, their evolution can be rather elaborate and even unpredictable. Wolfram has made an exhaustive study of one dimensional cellular automata, in which the cells are arranged on a line. Higher dimensional automata are useful in analysing biological processes; for example a two dimensional automaton can be used to investigate neurogenesis in a membrane of undifferentiated precursor cells.[6]

7.1.3   Percolation

Consider a spatial array of at least two dimensions, with cells able to take values of zero or one, signifying respectively 'impermeable' and 'permeable' to some agent migrating across the array and only able to move from a permeable site to a nearest neighbour that is also permeable. Let p be the probability that a cell has the value 1. If ones are sparse the mobility of the agent will be restricted to small isolated islands. The most important problem in this field is to determine the mean value of p at which an agent can

---

[6]Luthi et al.

span the entire array via its nearest neighbour connexions. This is so-called 'site percolation'.[7]

A possible approach to determine the critical value p_c is as follows: the probability that a single permeable cell on a square lattice is surrounded by impermeable ones (i.e. is a singlet) is pq^4, where $q = 1 - p$. Defining $n_s(p)$ to be the average number of s-clusters per cell, then we have $n_2(p) = 2p^2q^6$ for doublets, (the factor 2 arises because of the two possible perpendicular orientations of the doublet), $n_3(p) = 2p^3q^8 + 4p^3q^7$ for triplets (linear and bent), etc. If there are few permeable cells, $\sum_s sn_s(p) = p$; if there are many we can expect most of the particles to belong to an infinite (in the limit of an infinite array) cluster, hence $\sum_s sn_s(p) + P_\infty = p$, and the mean cluster size $S(p) = \sum_s s^2 n_s(p)/p$. If $S(p)$ is now expanded in powers of p, one will find that at a certain value of p the series diverges; this is when the infinite (spanning) cluster appears, and we can call the array 'fully connected'. The remarkable Galam-Mauger formula for isotropic lattices

$$p_c = a[(D - 1)(C - 1)]^{-b} \qquad (7.12)$$

where D is the dimension, C the connectivity of the array (i.e. the number of nearest neighbours of any cell), and a and b are constants with values 1.2868 and 0.6160 respectively, allows one to calculate the critical threshold for many different types of network.

7.2   Networks

The cellular automata considered above (§7.1.2) are examples of *regular* networks (of automata): the objects are arranged on a regular lattice and connected in an identical fashion with neighbours. Consider now a collection of objects (nodes or vertices) characterized by number, type and interconnexions (edges or links). Figure 7.2 represents an archetypical fragment of a network (graph). The connexions between nodes can be given by a matrix A whose elements a_{ij} give the strength of the connexion (in a Boolean network $a = 1$ or 0, repectively connexion present or absent) between nodes i and j.

We begin by considering their structural properties. Useful parameters of a network are: N, the number of nodes; $\bar{k}$, the average degree of each node (the number of other vertices to which a given vertex is joined); L, the network diameter (the smallest number of edges connecting a randomly

---

[7]In 'bond percolation' movement occurs along links joining nearest neighbours with probability p. Every bond process can be converted into a site one, but not every site process is a bond one.

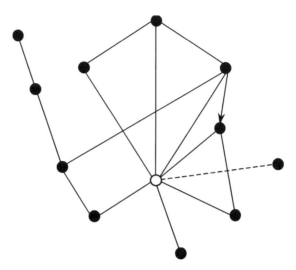

Figure 7.2: A fragment of a network (graph). Note the two types of nodes, and that some of the vertices are anisotropic.

chosen pair of nodes; this is a global property);[8] and the cliquishness $\mathfrak{C}$ defined as the fraction of nodes linked to a given vertex that are themselves connected (this is a local property), or in other words the (average) number of times any two nodes connected to a third node are themselves connected. Hence this is equivalent to the number of closed triangles in the network, i.e.

$$\mathfrak{C} \propto \mathrm{Tr}\, A^3 \, , \tag{7.13}$$

from which a relative clustering coefficient can be defined as

$$\mathfrak{C}_r = \mathfrak{C}/N \, . \tag{7.14}$$

The maximum number of possible edges in a network is $N(N-1)/2$ (the factor two arising because each edge has two endpoints).

Two important generic topologies of graphs are:

(i) *random or Erdős-Rényi graphs.* Each pair of nodes is connected with probability p: the connectivity of such a network peaks strongly at its average value and decays exponentially for large connectivities. The probability $p(k)$ that a node has k edges is given by $\mu^k e^{-\mu}/k!$, where $\mu = 2Np$ is the

---

[8]A useful way to compute L is given by Raine & Norris.

mean number of edges per node. The smallest number of edges connecting a randomly chosen pair of nodes (i.e. the network diameter L) is $\sim \log N$ (cf. $\sim N$ for a regular network). The cliquishness or clustering coefficient $\mathfrak{C} = \mu$.

This type of graph has a percolation-like transition. If there are M interconnexions, when $M = N/2$ a giant cluster of connected nodes will appear.

A special case of the random graph is the small world. This term applies to networks in which the smallest number of edges connecting a randomly chosen pair of nodes is comparable to the $\log N$ expected for a random network, i.e. much smaller than for a regular network, while the local properties are characteristic of a regular network, i.e. the clustering coefficient is high. The name comes from the typical response, "It's a small world!" uttered when it turns out that two people meeting for the first time and with no obvious connexion between them have a common friend.[9]

(ii) *the 'scale-free' networks*, in which the probability $P(k)$ of a node having k links $\sim k^{-\gamma}$, where γ is some constant. A characteristic feature of a scale free network is that it possesses a very small number of highly connected nodes. Many properties of the network are highly vunerable to the removal of these nodes.

Scale free networks appear to be widespread in the world. The first systematic investigation of their properties appears to have been conducted by Dominican monks in the 13th and 14th centuries, in connexion with eradicating heresy.

7.2.1 Trees

A tree is a graph in which each pair of vertices is joined by a unique edge; there is exactly one more vertex than the number of edges. In a binary tree each vertex has either one or three edges connected to it. A rooted tree has one particular node called the root. Trees represent ultrametric space satisfying the strong triangle inequality

$$d(x, z) \leq \max\{d(x, y), d(y, z)\} \tag{7.15}$$

---

[9]The first published account appears in F. Karinthy, *Láncszemek* (in: *Címszavak a Nagy Enciklopédiához*, vol. 1, pp. 349–354. Budapest: Szépirodalmi Könyvkiadó (1980). It was first published in the 1920s). A simple way of constructing a model small world network has been given by Watts and Strogatz: start with a ring of nodes each connected to their k-nearest neighbours (i.e. a regular network). Then detach connexions from one of their ends with probability p and reconnect the freed end to any other node (if $p = 1$ then we recover a random network). As p increases L falls quite rapidly, but $\mathfrak{C}$ only slowly (as $3(\mu - 2)/[4(\mu - 1)]$. The small world property applies to the régime with low L but high $\mathfrak{C}$.

where x, y and z are any three nodes and d is distance between a pair of nodes.

Trees are especially useful for representing hierarchical systems, in which elements are clustered according to the strengths of their interactions.

The complexity $\mathfrak{C}$ of a tree T consisting of b subtrees $T_1, \ldots, T_b$ (i.e. b is the number of branches at the root), of which k are not isomorphic, is defined as[10]

$$\mathfrak{C} = \mathfrak{D} - 1 \qquad (7.16)$$

where the diversity measure $\mathfrak{D}$ counts both interactions between subtrees and within them, and is given by

$$\mathfrak{D} = (2^k - 1) \prod_{j=1}^{k} \mathfrak{D}(T_j^{(i)}) . \qquad (7.17)$$

If a tree has no subtrees, $\mathfrak{D} = 1$; the complexity of this, the simplest kind of tree, is set to zero (hence equation 7.16). Any tree with a constant branching ratio at each mode will also have $\mathfrak{D} = 1$ and hence zero complexity. This complexity measure satisfies the intuitive notion that the most complex structures are intermediate between regular and random ones.

7.2.2   Complexity parameters

There are various measures of network complexity:

1. κ, the number of different spanning trees of the network;

2. Structural complexity, the number of parameters needed to define the graph;

3. Edge complexity, the variability of the second shortest path between two nodes;

4. Network or β-complexity, given by the ratio $\mathfrak{C}/L$;

5. Algorithmic complexity, the length of the shortest algorithm needed to describe the network (see also Chapter 6).

---

[10]See Huberman & Hogg.

7.2.3 Dynamical properties

The essential concepts of physical structure and state structure were already introduced above (§7.1.1 and figure 7.1). A considerable body of work has been accomplished along these lines: investigating the state structures of simple, or simply constructed, networks. Kauffman in particular has studied large randomly connected Boolean networks, with the interesting result that if each node has on average two inputs from other nodes, typically the state structure comprises about $N^{1/2}$ cyclic attractors, where N is the number of nodes, i.e. far fewer than the 2^N potentially accessible states.

More generally, Kauffman considers strings of N genes, each present in the form of either of two alleles (0 and 1).[11] In the simplest case each gene is independent, and when a gene is changed from one allele to the other, the total fitness changes by at most $1/N$. If epistatic interactions are allowed, the fitness contribution depends on the gene plus the contributions from K other genes (the NK model), and the fitness function or 'landscape' becomes less correlated and more rugged.

Érdi & Barna have studied how the pattern of connexions changes when their evolution is subjected to certain simple rules; the evolution of networks of automata in which the properties of the automata themselves can change has barely been touched, although this, the most complex and difficult case, clearly is the one closest to natural networks within cells and organisms. The study of networks and their application to real world problems has, in fact, only just begun.

7.3 Synergetics

General systems theory (§7.1) can be further generalized and made more powerful by including a diffusion term:

$$\frac{\partial u_i}{\partial t} = \frac{1}{\tau_i} F_i(u_1, u_2, \ldots, u_n) + D_i \Delta u_i, \qquad i = 1, 2, \ldots, n . \qquad (7.18)$$

u_i is a dynamic variable, e.g. the concentration of the ith object at a certain point in space, $F_i(u_i)$ are functions describing the interactions, τ_i is the characteristic time of change, and D_i is the diffusion coefficient (diffusivity) of the ith object. (7.18) is thus a reaction-diffusion equation that explicitly describes the spatial distribution of the objects under consideration. The diffusion term tends to zero if the diffusion length $l_i > L$, the spatial extent of the system, where

$$l_i = D_i^{1/2} \tau_i . \qquad (7.19)$$

---

[11]Here we preempt some of the discussion in §9.8.1.

Although solutions of equation (7.18) might be difficult for any given case under explicit consideration, in principle we can use it to describe any system of interest. This area of knowledge is called synergetics. Note that the "unexpected" phenomena, which are often observed in elaborate systems, can be easily understood within this framework, as we shall see.

One expects that the evolution of a system is completely described by its n equations of the type (7.18), together with the starting and boundary conditions. Suppose that a stationary state has been reached, at which all the derivatives are zero, and described by the variables $\bar{u}_1, \ldots, \bar{u}_n$ at which all the functions F_i are zero. Small deviations δu_i may nevertheless occur, and can be described by a system of linear differential equations

$$\frac{\mathrm{d}}{\mathrm{d}t}\delta u_i = \sum_{j}^{n} a_{ij}\delta u_j \tag{7.20}$$

where the coefficients a_{ij} are defined by

$$a_{ij} = \left.\frac{\partial F_i}{\partial u_i}\right|_{u_i=\bar{u}_i}. \tag{7.21}$$

The solutions of (7.20) are of the form

$$\delta u_j(t) = \sum_{j}^{n} \epsilon_{ij} e^{\lambda_i t} \tag{7.22}$$

where the ϵ_{ij} are coefficients propotional to the starting deviations, viz. $\epsilon = \delta u(0)$. The λs are called the Lyapunov numbers, which can in general be complex numbers, the eigenvalues of the system: they are the solutions of the algebraic equations

$$\det|a_{ij} - \delta_{ij}\lambda_j| = 0 \tag{7.23}$$

where δ_{ij} is Kronecker's delta.[12] We emphasize that the Lyapunov numbers are purely characteristic of the system, i.e. they are not dependent on the starting conditions or other external parameters—provided the external influences remain small.

If all the Lyapunov numbers are negative, the system is stable—the small deviations decrease in time. On the other hand, if at least one Lyapunov number is positive (or, in the case of a time-dependent Lyapunov number, if the real part becomes positive as time increases), the system is unstable, the deviations increase in time, and this is what gives rise to "unexpected" phenomena. If none are positive, but there are some zero or pure imaginary ones, then the stationary state is neutral.

---

[12] $\delta_{ij} = 0$ when $i \neq j$ and 1 when $i = j$.

7.3.1 Some examples

The simplest bistable system is described by

$$\frac{du}{dt} = u - u^3 . \tag{7.24}$$

There are three stationary states, at $u = 0$ (unstable; the Lyapunov number is $+1$) and $u = \pm 1$ (both stable), for which the equation for small deviations is

$$\frac{d}{dt}\delta u = -3\delta u \tag{7.25}$$

and the Lyapunov numbers are -3. This system can be considered as a memory box with an information volume equal to $\log_2(\text{no stationary states})$ $= 1$ bit.

A slightly more complex system is described by the two equations

$$\left.\begin{array}{l} du_1/dt = u_1 - u_1 u_2 - au_1^2 \\ du_2/dt = u_2 - u_1 u_2 - au_2^2 \end{array}\right\} . \tag{7.26}$$

The behaviour of such systems can be clearly and conveniently visualized using a phase portrait (e.g. figure 7.3). To construct it, one starts with arbitrary points in the (u_1, u_2) plane and uses the right hand side of equation (7.26) to determine the increments. The main isoclines (at whose intersections the stationary states are found) are given by

$$\left.\begin{array}{l} du_1/dt = F_1(u_1, u_2) = 0 \\ du_2/dt = F_2(u_1, u_2) = 0 \end{array}\right\} . \tag{7.27}$$

Total instability, in which every Lyapunov number is positive, results in dynamic chaos. Intermediate systems have strange attractors (which can be thought of as stationary states smeared out over a region of phase space rather than contracted to a point), in which the chaotic regime occurs only in some portions of phase space.

7.3.2 Reception and generation of information

If the external conditions are such that in the preceding example (equation 7.26) the starting conditions are not symmetric, then the system will ineluctably arrive at one of the stationary states, as fixed by the actual asymmetry in the starting conditions. *Hence information is received.*

On the other hand if the starting conditions are symmetrical (the system starts out on the separatrix), the subsequent evolution is not predetermined

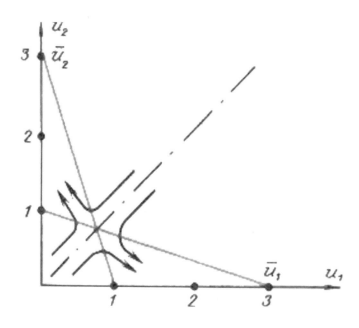

Figure 7.3: Phase portrait of the system represented by equation (7.26) with $a = 1/3$. The main isoclines (cf. 7.26) are $u_1 = 0$ and $u_2 = 1 - au_1$ ('vertical', determined from $F_1 = u_1 - u_1u_2 - au_1^2 = 0$ with $\Delta u_1 = 0$), and $u_2 = 0$ and $u_1 = 1 - au_2$ ('horizontal', determined from $F_2 = u_2 - u_1u_2 - au_2^2 = 0$ with $\Delta u_2 = 0$). The system has four stationary states: at $u_1 = u_2 = 0$, unstable, $\lambda_1 = \lambda_2 = +1$; at $u_1 = u_2 = 1/(1 + a)$, unstable (saddle point), $\lambda_1 = -1, \lambda_2 = (1 - a)/(1 + a) > 0$; at $u_1 = 1/a, u_2 = 0$, stable, $\lambda < 0$; and at $u_2 = 1/a, u_1 = 0$, stable, $\lambda < 0$. The separatrix (separating the basins of attraction) is shown by the dashed-dotted line (after Chernavsky).

and the ultimate choice of stationary state occurs by chance. *Hence information is generated.*[13]

7.3.3 Evolutionary systems

Equilibrium models, which are traditionally often used to model systems, are characterized by the following assumptions:

1. Microscopic events occur at their average rate;

2. Entities of a given type are identical, or their characteristics are normally distributed around a well-defined mean;

3. The system will move rapidly to a stationary (equilibrium) state (this movement is enhanced if all agents are assumed to perfectly anticipate what the others will do).

Hence only simultaneous, not dynamical, equations need be considered, and the effect of any change can be evaluated by comparing the stationary states before and after the change.

The next level in sophistication is reached by abandoning assumption 3. Now several stationary states may be possible, including cyclical and chaotic ones (strange attractors).

If only assumption 2. is kept, non-average fluctuations are permitted, and behaviour becomes much richer. In particular, external noise may allow the system to cross separatrices. The system is then enabled to adopt new régimes of behaviour, exploring regions of phase space inaccessible to the lower level systems,[14] which can be seen as a kind of collective adaptive response (requiring noise) to changing external conditions.

The fourth and most sophisticated level is achieved by abandoning the remaining assumption, 2. Local dynamics cause the microdiversity of the entities themselves to change. Certain attributes may be selected by the system and others may disappear. These systems are called evolutionary. Their structures reorganize, and the equations themselves may change. Most natural systems seem to belong to this category. Rational prediction of their future is extremely difficult.

---

[13]Cf. the discussion in Chapter 2.
[14]This type of behaviour is sometimes called "self-organization".

Part II

Biology

Chapter 8

Introduction to Part II

The primary purpose of the next two chapters is to give an overview of living systems, especially directed at the bioinformatician who has previously dealt purely with the computational aspects of the subject.

Whenever confronting the totality of biology, it is clear that one may approach it at various levels—molecular, cellular, organismal, populational, ecological. Traditionally these levels have been accorded official status by naming departments after them. Just as we saw with the levels of information (technical, semantic, effective), however, one quickly distorts a vision reflecting reality by insisting on the independence of these levels. For example, it it not possible to understand how populations of organisms evolve without considering what is happening to their DNA molecules. When reading the two following chapters, this interdependence should constantly be borne in mind.

Problem. Attempt to provide a definition of life. Find exceptions.

8.1 Genotype, phenotype and species

The basic unit of life is the organism. The phenotype may be defined as the organism interacting with its environment. The genotype may be defined as the set of instructions for the self-reproduction of the organism, and is supposed to be barely influenced by external conditions.

Those adhering to the primacy of the genome will nevertheless concede that sending the complete gene sequence of an organism to an alien civilization will not allow them to reconstruct it, i.e. create a living version of the organism, i.e. its phenotype. Many things, including the principles of chemical catalysis necessary for the genetic instructions to be read and processed, are not represented and not even implicit in the nucleic acid sequence. In fact, the phenotype is a composite of implicit and explicit meaning; the latter is context-dependent.

It must also be considered that the processes of natural selection, to be considered more deeply in Chapter 9, operate on the phenotype, yet the vehicle for their persistence is the genotype.

Organisms are characterized as species. Despite the pervasive use of the term in biology, no entirely satisfactory definition of 'species' exists. 'Reproductive isolation' is probably one of the better operational definitions, but it can only apply under carefully circumscribed conditions. Geographical as well as genetic factors play a rôle, and epigenetic factors are even more important. In any human settlement of at least moderate size, there are almost certainly groups of inhabitants with no social contact with other groups at all: hence these groups are as effectively reproductively isolated from each other, because of behavioural patterns, as if they were living on different continents, and if we apply our definition, we are forced to assert that the groups belong to different species (even though both are taxonomically classified as *Homo sapiens*).

The concept of reproductive isolation is of little use when species reproduce asexually (such as bacteria); in this case a criterion based on the possibility of significant exchange of genetic material with other organisms has to be used.

Another difficulty in defining 'species' in terms of associating them with autonomously reproducing DNA is that not only are there well-defined organisms such as coral or lichen in which two 'species' are actually living together in inseparable symbiosis, but we ourselves host about 10^{14} unicellular organisms, which comfortably outnumber the 10^{13} or so of our own cells.

A very striking characteristic of living organisms is that they are able to maintain their being in changing surroundings. It is doubtful whether any artificial machine can survive over as wide a range of conditions as man, for example. 'Survival' means that the essential variables of the organism are maintained within certain limits. This maintenance (homeostasis) requires regulation of the vital processes. We shall consider regulation more formally in the following section.

The little table below puts some of the terms encountered into a kind of correspondence. It is not a table of synonyms.

Problem.   Discuss and extend table 8.1. Find descriptive headings for the columns.

genotype	phenotype
genetics	epigenetics
nature	nurture
gene	environment
necessity	chance (or freedom)
K	I (from equation (2.12))
explicit	implicit
syntax	semantics
. . .	. . .

Table 8.1: Table of approximate equivalents. See text.

8.2 Regulation

Regulation may be considered in abstract terms common to any mechanism, whether living or not. The formalism presented below will be explicitly made use of in Chapter 14 when considering signalling and regulatory pathways.

The essential elements of a regulatory system are shown in figure 8.1. The lines connecting the components indicate communication channels. The dotted lines indicate the paths along which the regulator can receive information about the disturbance. By way of illustration, consider the operation of a simple thermostatted water bath. T then represents the electric heater and the bath itself with a circulator. E represents the water temperature (measured with a thermometer) T, R the switch controlling the power supplied to the heater, and D the disturbances from the environment. A typical event is the immersion of a flask containing liquid at a temperature lower than that of the bath. Sophisticated baths may be able to sense the temperature and mass of the flask before it has been immersed (channel D → R), or at the moment of its placement (channel T → R), but typically the heater is switched on if the temperature falls below the target value T_0 (channel E → R). This is called regulation by error. Most living cells appear to operate according to this principle.

The canonical representation of the thermostat is

$$\downarrow \begin{array}{cc} a & b \\ a & a \end{array} \qquad (8.1)$$

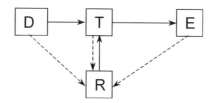

Figure 8.1: Schematic diagram of a regulatory mechanism. The components are: D, disturbance (from the environment); T, the hardware or the mechanism; R, the regulator; and E, the essential variables (output). See the text for further explanation.

where state a represents $T = T_0$ (within the allowed uncertainty), and state b represents $T < T_0$. In the case of a bacterium, a may represent $[\mathrm{Hg}^{2+}] = 0$ (square brackets denoting concentration), and b $[\mathrm{Hg}^{2+}] > 0$. D in Fig. 8.1 now corresponds to mercury ions in the environment of the cell, T to the proteins able to sense mercury ions and the gene expression machinery able to synthesize mercury reductase, and R to the transcription factor binding to the mercury reductase gene promoter sequence. In stochastic matrix representation, we have

$$
\begin{array}{c|cc}
\downarrow & a & b \\
\hline
a & 1 & 1 \\
b & 0 & 0
\end{array}
\qquad . \tag{8.2}
$$

More realistically, however, we might have

$$
\begin{array}{c|cc}
\downarrow & a & b \\
\hline
a & 1 & 0.6 \\
b & 0 & 0.4
\end{array}
\tag{8.3}
$$

for example, since for various reasons the machinery may not work perfectly. Further sophistication may be incorporated by increasing the number of states, e.g. a, b, c, d corresponding respectively (for example) to $[\mathrm{Hg}^{2+}] =$

0, 1 nM, 1μM, 1 mM and above, with the corresponding matrix

$\downarrow$	a	b	c	d
a	1	0.6	0.3	0
b	0	0.4	0.4	0.3
c	0	0	0.3	0.4
d	0	0	0	0.3

After several cycles, the machine will be completely in state a (cf. §6.2).

In the simplest cases, the error, or a quantity proportional to it, is sent back to the regulator, but more sophisticatedly some function of the error, e.g. its integral, or its derivative, could be fed back to R. The vast majority of industrial controllers use a combination of all three (and hence are referred to as PID controllers).

8.3   The concept of machine

'Machine' is used formally to describe the embodiment of a transformation (e.g. 8.1). The essential feature is that the internal state of the machine, together with the state of its surroundings, uniquely defines the next state to which it will go. A determinate machine is canonically represented by a closed, single valued transformation (equations 8.1 and 8.2); a Markovian machine is indeterminate insofar as the transitions are governed by a stochastic matrix (e.g. equation 8.3); the determinate machine is clearly a special case of the more general Markovian machine.

If there are several possible transformations, and a parameter governs which transformation shall be applied to the internal states of the machine, then we can speak of a machine with input, the input being the parameter. The machine with input is therefore a transducer.

A Markovian machine with input would be represented by a set of stochastic atrices together with a parameter to indicate which matrix is to be applied at any particular step.

8.4   Adaptation

The process of adaptation, so ubiquitous in nature, has been formalized by Sommerhoff. The 'disturbance' introduced above is denoted the *coenetic variable*, the 'hardware' (T) is the environmental circumstance E_{t_1}, the 'regulation' (R) is the response R_{t_1} directively correlated with E_{t_1}, and the 'essential variables' are the focal condition.

The usual notion of adaptedness, as applied to biological systems, implies no more than appropriateness. In other words, the statement that an (organic) response R is adapted to the environmental circumstances E from the viewpoint of some future state of affairs F (towards the realization of which it is conceived to be directed), implies that the response is appropriate, and hence also effective in bringing about the actual (or probable at least) occurrence of F.

This 'definition' of adaptedness, although easy to state, is not only trivial, but is also fraught with difficulties. For one thing, it does not allow us to prefer the statement 'an aquarium is adapted to the fish it contains' to 'the fish is adapted to the aquarium in which it survives'. Another difficulty is presented by that numerous category of accidental activity. Many accidental occurrences (including random mutations of DNA) are highly effective in bringing about a certain reponse, but could hardly be called adapted—in the case of a mutation, adaptation could be said to have occurred only after it had become fixed in the population due to the advantages it conferred on the organism.

Sommerhoff's formulation begins with the recognition that adaptedness means that a response R is adapted with respect to F if the mechanism engendering it is such that were the initial circumstance E different (taken from a defined set E_1, E_2, E_3, etc.) then the reponse would have been, correspondingly, R_1, R_2, R_3, etc., i.e. there is a systematic correlation between the sets E and R, and this systematic correlation is called *directive correlation*. In other words, the coenetic variable is the stimulus, or particular initial environmental condition, which at time t jointly causes the particular response R and enacts the condition E; directive correlation is the response R, directively correlated to the simultaneous conditions E in respect of the future state of affairs F if the system of which they are part is so conditioned that there exists an event prior to t such that the corresponding variation of the coenetic variable implies variations of both R and E, and the correlated sets of possible values of R and E are such that all pairs of corresponding members of the sets, but no others, cause the subsequent occurrence of F; and the focal condition is the formalized future state of affairs, F.

Chapter 9

The nature of living things

Figure 9.1 shows, in highly compressed and schematic form, the major processes taking place within living beings. The first priority of any living being is simply to survive. In the language of §8.2, the being must maintain its essential variables within the range corresponding to life. In most succint form, 'to be or not to be—that is the question'.

The biosynthetic processes of life maintenance, indicated at the bottom of the figure, lead beyond the living part of the organism to produce external structures, like exoskeletons and shells, but sometimes huge, such as coral reefs, treetrunks, guano hills, and indeed beaver dams and buildings of human construction.

Bioinformatics is particularly concerned with the processes of information flow (cf. the 'central dogma'), i.e. d, e, f, and regulation of those processes (g, h, i). Nevertheless, any student of bioinformatics should have some grasp of the overall picture, which this chapter sets out to give.

The simplest organisms are single cells, slightly more elaborate organisms such as sponges consist of aggregates of cells constrained to live together, and more complex organisms are highly constrained assemblies of cells.

9.1 The cell

The basic unit of life is the cell. Many organisms consist of only one cell. Therefore even a single cell carries all that is needed for life. The cell contains the DNA coding for proteins and all the machinery necessary for maintaining life—enzymes, multiprotein complexes, etc. The body of the cell, the cytoplasm, is a thick, viscous aqueous medium full of macromolecules. If intact cells are centrifuged, one can separate a fairly fluid fraction, which contains

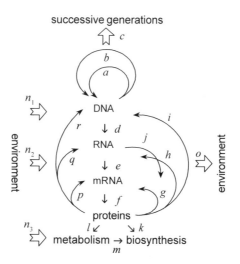

Figure 9.1: Schematic diagram of the major relations in living organisms. The innermost zone concerns processes taking place within the cell. The upper portion indicates processes (a, b) involved in multiplication (reproduction); the lower portion, processes (k, l, m) involved in life maintenance (homeostasis). The curved arrows moving upwards in the central region on the left hand side indicate processes (p, q, r) of synthesis; those on the right hand side (g, h, i, j) processes of regulation. Exchange with the environment (input and output) takes place: inputs n_1, n_2, n_3 could be, respectively, cosmic rays causing DNA mutations, toxicants interfering with the regulation of transcription and translation, and food. The successive generations (c) are of course released into the environment.

very little apart from a few ions and small osmolytes like sugars.[1] The rest of the proteins etc. which are usually called "cytoplasmic" are bound to macromolecular constructs such as the inner surface of the outer lipid membrane, internal membranes, other polymers such as various filaments (the cytoskeleton) made from proteins such as actin, tubulin etc., or polysaccharides. These bound proteins can only be released if the ultrastructure of the cell is completely disrupted, e.g. by mechanically crushing it in a cylinder in which a tightly fitting piston moves; the results obtained from fractionating such homogenates give a quite misleading impression of the constitution of a living cell.

The cell membrane (also called 'plasma membrane' or 'plasmalemma'), often described as a robust and fairly impermeable coating around the cytoplasm, has a function which strictly speaking remains somewhat mysterious, since modern, and not so modern, research has shown that cells remain viable even when their membranes are significantly disrupted. The image of a cell as a toy balloon filled with salt solution, which would spurt out if the balloon were punctured, is not in agreement with the experimental facts.

9.1.1 The structure of a cell

The two great divisions of cell types are the prokaryotes (bacteria and archaea) and the eukaryotes (protozoa, fungi, plants and animals) (see §9.9 appended to this chapter). As the name suggests, the eukaryotes possess a definite nucleus containing the genetic material (DNA), and separated from the rest of the cell by a lipid-based membrane, whereas the prokaryotes do not have this internal compartmentation. Moreover the eukaryotes possess other internal compartments known as organelles: the mitochondria, sites of oxidative reactions where food is metabolized; chloroplasts (only in plants), sites of photosynthesis; lysosomes, sacs of digestive enzymes for decomposing large molecules; the endoplasmic reticulum, a highly folded and convoluted lipid membrane structure to which the ribosomes, RNA-protein complexes responsible for protein synthesis from mRNA templates are attached, and contiguous with the Golgi body, responsible for other membrane operations such as packaging proteins for excretion to outside the cell; etc. The mitochondria and chloroplasts possess their own DNA, which codes for some, but not all their proteins; they are believed to be vestiges of formerly symbiotic prokaryotes living within the larger eukaryotes. The present interrelationship between cell and mitochondrion is highly convoluted. The yeast mitochondrion, for example, has about 750 proteins, of which only 8 are templated by

---

[1]Kempner and Miller.

the mitochondrial genome, the remainder coming from the principle genome of the cell.

Observational overview

The optical microscope can resolve objects down to a few hundred nanometres in size. This is sufficient for revealing the existence of individual cells (Hooke, 1665) and some of the larger organelles (subcellular organs) present in eukaryotes. The contrast of most of this internal structure is low, however, and stains must be applied in order to clearly reveal them. Thus, nucleus, chromosomes, mitochondria, chloroplasts etc. can be discerned, even though their internal structure can not. The electron microscope, capable of resolving structures down to sub-nanometre resolution, has vastly increased our knowledge of the cell, although it must always be borne in mind that the price of achieving this resolution is that the cell has to be killed, sectioned, dehydrated or frozen, and stained or fixed, procedures which are susceptible to alter many of the structures from their living state.[2] Mainly through electron microscopy, a large number of structures, such as microfilaments, microtubules, endoplasmic reticulum, Golgi bodies, lysosomes, peroxysomes etc. acquired something more than a shadowy existence.

If cells are mechanically homogenized, different fractions can be separated in the centrifuge: lipid membrane fragments, nucleic acids, proteins, polysaccharides and a clear, mobile aqueous supernatant containing small ions and osmolytes. It should not be supposed that this supernatant is representative of the cytosol, the term applied to the medium surrounding the subcellular structures, in the living cell: centrifugation of intact cells (the experiments of Kempner and Miller) removes practically all macromolecules along with the lipid-based structures. That experiment was done relatively late in the development of biochemistry, after the misconception that the cytosol was filled with soluble enzymes had already become established. Most proteins are attached to membranes, and the cytosol is a highly crowded, viscous hydrogel.

Lipid membranes occupy a very important place in the cell. Most of the organelles are membrane-bounded, and their surfaces are the sites of most enzyme activity. Chloroplasts are virtually filled with internal membranes. Curiously, the most prominent membrane of all, that surrounding the cell, has even today a rather obscure function: it is often maintained, for example, that it is needed to control ion fluxes into and out of the cell, but experimentally, potassium flux is unaffected by removal of the membrane (Solomon, 1960).

---

[2]See Hillmann for an extended discussion.

Although prokaryotes (which are mostly much smaller than eukaryotes) lack most of the internal membrane-based structure seen in eukaryotes, they are still highly heterogeneous in terms of the distributions of components, from macromolecules down to small ions, which are highly nonuniform.

If molecules are tagged, by synthesizing them with unusual isotopes, or attaching a fluorescent label, or a nanoparticle, individual molecules, or small groups of molecules, can be localized in the cell, by spatially resolved secondary ion mass spectrometry (SIMS), fluorescence microscopy, etc. These measurements can usually be carried out with fair time resolution (ms–s), hence both local concentrations and fluxes of these molecules can be determined.

Spontaneous assembly. Take the isolated constituents of a phage virus, mix them together, and a functional virus will result. This exercise cannot be repeated successfully with larger, more complex structures closer to that state we call "living". Nor does it work if we break down the phage constituents into individual molecules.

9.2 The cell cycle

Just as exponential decay is an archetypical feature of radioactivity, so is exponential growth an archetypical feature of the observable characteristics of life. If a single bacterium is placed in rich nutrient medium, after a while (as little as twenty minutes in the case of *Escherichia coli*) two bacteria will be observed; after another twenty minutes, four, and so on, i.e. the number n of bacteria increases with time t as e^t (cf. equation 7.4).

Actually exponential growth, as known under laboratory conditions, is not very common in nature. The vast majority of bacteria in soils and sediments live a quiet, almost moribund existence, due to the scarcity of nutrient. Under transiently favourable conditions, growth might start out exponentially, but would then level off as nutrients became exhausted (cf. equation 7.5).

Bacteria 'multiply by division'. Since the average size of each individual bacterium remains roughly constant averaged over long intervals, what actually happens is that the first bacterium increases in size, and then divides into two. The division does not appear to be symmetrical in general, in other words, to express the result of the division as 'two daughter cells' may not be accurate; there is a mother and daughter, and they are not equivalent.[3]

---

[3]The events of growth and division are not really akin to printing multiple copies of a book, or photocopying pages. It is not strictly speaking correct to call the process whereby adult organisms create new organisms—offspring—'reproduction': parents do not

During the growth process, most of the molecules of the cell are increasing (in number) *pro rata*, including the cell's gene, a circle of double stranded DNA. Once the gene has been duplicated, the rest of the material can be divided, and growth starts again. The process has a cyclic nature, and is called the cell cycle (figure 9.2).

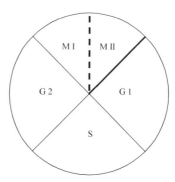

Figure 9.2: Schematic diagram of the cell cycle. The successive epochs are known as 'phases' (cf. phases of the moon). Areas of the sectors are proportional to the typical duration of each phase. The phases succeed each other in a clockwise direction. A newly born cell starts in the so-called G1 phase. When it reaches a certain size (the molecular nature of the initiating signal is not known, but it is correlated with size) DNA synthesis begins, i.e. the gene is duplicated. Mitosis (see below) takes place in the M phase. See also table 9.1.

The defining events are: initiation of chromosome replication, chromosome segregation, cell division, and inactivation of the replication machinery. The duration of one cycle can vary by many orders of magnitude: 20 minutes for *E. coli* grown in the laboratory, to several years for the bacteria believed to live in deep ocean sediments. Many fully differentiated cells never divide.

Apart from duplicating its DNA and dividing, the cell also has to metabolize food (to provide energy for its other activities, which may be secreting certain substances etc., or simply playing a structural rôle), and neutralize external threats such as viruses, toxins, changes in temperature etc. All these activities, including gene duplication, require enzymes, and enzymes for

---

reproduce themselves when they make a baby; even when the baby is grown up it might be quite different, in appearance and behaviour, from its progenitors. In a literary analogy, this kind of process is akin to writing a new book (but a derivative work) by gathering material from primary sources, or previously existing secondary sources.

Phase	process	feature(s)
M	prophase	chromosome condensation
M	metaphase	centrosomes separate and form two asteriated poles at opposite ends of the cell
M	prometaphase	nuclear envelope is degraded, microtubules from the centrosomes seek the chromosomes
M	metaphase	microtubules from the centrosomes find the chromosomes
M	anaphase A	the two arms of each chromosome are separated and drawn towards the centrosomes
M	anaphase B	centrosomes move further away from each other together with their half-chromosomes
M	telophase	the cell divides
G1	decondensation	chromosomes disappear, nuclear envelope reforms around the DNA, microtubules reappear throughout the cytoplasm
S	interphase	cell growth
G2	interphase	DNA duplication

Table 9.1: Successive events in the eukaryotic cell cycle. The nuclear envelope is a bilayer lipid membrane in which proteins are embedded. The centrosomes are large multiprotein complexes. Mitosis (see 9.2.1) is considered to begin at the end of G2, and last until the beginning of G1.

translating and modifying the nucleic acid genetic material, whose fabrication also requires energy. There is also a considerable amount of degradation activity, i.e. proteolysis of enzymes after they have carried out their specific function.[4] Degradation of course itself requires enzymes to carry it out. In eukaryotes, most proteins are marked for degradation by being covalently bound to one or more copies of the polypeptide ubiquitin. This facilitates their recognition by a huge ($M_r \sim 10^6$) multiprotein complex called the proteasome, which carries out the proteolysis into peptides, which may be presented to the immune system, and ultimately to amino acids.

9.2.1   The chromosome

In eukaryotes, the nucleic acid is present as long linear segments, each containing thousands of genes, called chromosomes, because they can be coloured (stained), and hence rendered visible in the optical microscope during cell division.

Chromosomes are terminated by telomeres. The telomere is a stretch of highly repetitive DNA. Since during chromosome replication (see below) the DNA polymerase complex stops several hundred bases before the end, telomeres prevent the loss of possibly useful genetic information.

Germline cells are haploid, i.e. they contain one set of genes (like bacteria). When male and female gametes (eukaryotic germline cells) fuse together, the zygote, the single-celled progenitor of the adult organism) therefore contains two sets of genes (i.e. two double helices), one from the male and one from the female parent. This state is called diploid. The normal descendents of the zygote, produced by mitosis, remain diploid. Many plants, and a few animals, have more than two sets (four = tetraploid, many = polyploid), widening the possibilities for the regulation of gene expression. Polyploidy is a macromutation that greatly alters the biochemical, physiological and developmental characteristics of organisms. It may confer advantageous tolerance to environmental exigency (especially important to plants, because of their immobility), and open new developmental pathways.

The two (or more) forms of the same gene are called alleles. The inheritance of unlinked genes (i.e. genes on different chromosomes; genetic linkage refers to the association of genes located on the same chromosome) follows Mendel's laws.[5] If there are two alleles known for a given gene, denoted A

---

[4]A good example of this kind of enzyme is cyclin, which has a regulatory function, and whose concentration rises and then falls during mitosis.

[5]1. Phenotypical characters depend on genes. Each gene can vary, the ensemble of variants being known as alleles. In species reproducing sexually, each new individual receives one allele from the father and one from the mother. 2. When an individual

and a, occurring with probabilities p and $1 - p = q$ respectively, there are three possible genotypes in the population, AA Aa and aa, with probabilities of occurrence of p^2, $2pq$ and q^2 respectively (the Hardy-Weinberg rule). The Aa genotype is called heterozygous (the two parental alleles of a gene are different).

The existence of a maternal and a paternal gene is typical of eukaryotes, i.e. brothers and sisters share half their genes with each other. The social insects are an important recall that about a quarter of the animal mass on earth comprises ants) exception. The queen is only fertilized once in her lifetime, storing the sperm in her body. She lays two kinds of eggs, fertilized with the stored sperm, just before laying, which become females, and unfertilized, which become males. The males therefore have only one set of chromosomes, i.e. they are haploid. In a sense, the males have no father. Hence they transmit all their genes to their progeny, which are invariably female In consequence, sisters share three quarters of their genes with each other, but only have a quarter of their genes in common with their brothers.

Differences between prokaryotes and eukaryotes

Prokaryotes do not undergo meiosis, nor mitosis (their DNA is segregated as it replicates), their chromosomes are not organized into chromatin, nor does the DNA spend much of its time inside a special compartment, the nucleus (although the chromosome is usually visible as a compact body called the nucleoid). Chromosome replication typically starts from a single site in prokaryotes (the origin of replication, *ori*, which may comprise a few hundred bases), but from many (thousands) in eukaryotes—otherwise replication, proceeding at about 50 bases/s, would simply take far too long. As it is, the human genome takes about eight hours to be replicated. Prokaryotic DNA is circular (and hence does not require telomeres), whereas eukaryotic DNA is linear.

Mitosis

The simple process of gene replication is called mitosis. This is the type of cell division that produces two genetically identical (in theory) cells from a single parent cell. It applies to bacteria and to the somatic (body) cells of eukaryotes.

---

reproduces, it transmits to each offspring the paternal allele with probability 1/2 and the maternal allele with probability 1/2. 3. The actual transmission events are independent for each independently conceived offspring.

Prior to division, homologous pairs (of the maternal and corresponding paternal gene for each chromosome) form. They are attached at one zone, near the centre of the chromosome, by a protein complex called the centromere. The attached chromosomes then compactify, forming the characteristic 'X' shaped structures easily seen in the optical microscope after staining. The remainder of the process is described in §9.1.

Meiosis

Meiosis is a more complex process than mitosis. It starts with an ordinary diploid cell and leads to the formation of gametes (germline cells).

Firstly the two chromosomes (paternal and maternal) are duplicated (as in mitosis) to produce four double helices. Then the four double helices come into close proximity and recombination (see below) is possible. Thereupon the cell divides without further DNA replication. The chromosomes are segregated, hence each cell contains two double helices (diploid). A given double helix may have sections from the father and from the mother. Finally there is a further division without further DNA replication. Each cell contains one double helix (haploid). They are the gametes (germ cells).

9.2.2   The structure of genes and genome

Definition.   'Gene' is defined as a stretch of DNA that codes for (i.e. is translated into—see §9.5—a protein.[6]

Definition.   The genome is defined as the entire set of genes in the cell. Intergenomic sequences and introns (a term suggested by Walter Gilbert in 1978, signifying intragenic sequences) were not known when the word was coined. It therefore includes all polymerized nucleic acids.

The most basic genome parameter is the the number of bases (base pairs, since most genetic DNA is double stranded). Sometimes the molecular weight of the DNA is given (the average molecular weight of the four base pairs is 660). Table 9.2 gives the sizes of the genomes of some representative organisms.

Bacterial genomes consist of blocks of genes preceded by regulatory (promoter) sequences. Eukaryotic DNA is a mosaic of genes (segments whose sequence codes for amino acids, also called exons, from *ex*pressed); segments

---

[6]Formerly, the term 'cistron' was used to denote the genetic unit of function corresponding to one polypeptide chain. The discovery of introns (see below) signified the end of the 'one gene, one enzyme' idea.

Organism	No bp	No genes	No chromosomes	No cell types (approx.)
Escherichia coli	4×10^6	4290	1	1
Streptomyces coelicolor	8.6×10^6	7830	1	2
Amoeba dubia	7×10^{11}	?	~ 300	1
S. cerevisiae	10^7	6300	16	2
C. elegans	9×10^7	19 000	6	30
D. melanogaster	1.8×10^8	13 500	8	50
Oikopleura dioica	7×10^7	15 000	?	?
Protopterus (lungfish)	1.4×10^{11}	?	38	?
Triturus (newt)	1.9×10^{10}	?	?	150
Mus musculus	3.5×10^9	30 000	20	?
Homo sapiens	3.5×10^9	30 000	23	220
Neurospora crassa	4×10^7	10 000	~ 1000	?
Ophioglossum (fern)	?	?	~ 1000	?
Arabidopsis thaliana	1.3×10^8	25 500	10	?
Fritillaria	1.3×10^{11}	?	?	?

Table 9.2: Typical genome data.

(called introns) which are transcribed into RNA, but then excised to form the final mRNA used as the template for producing the protein (many genes are split into a dozen or more segments, which can be spliced in different ways to generate variant proteins after translation); promoters (short regions of DNA to which RNA, proteins or small molecules may bind, modulating the attachment of RNA polymerase to the start of a gene); and intergenomic sequences (the rest, sometimes called "junk" DNA in the same sense in which untranslated cuneiform tablets may be called junk—we do not know what they mean). This is schematically illustrated in figure 9.3.

Although the DNA to protein processing apparatus involves much complicated molecular machinery, some RNA sequences can splice themselves. This autosplicing capability enables exon shuffling to take place, suggesting the combinatorial assembly of exons as irreducible codewords as the basis of primitive, evolving life.

Organisms vary enormously in the proportion of their genome which is not genes. The intergenomic material may exceed by more than an order of magnitude the quantity of "coding" DNA. Some of the intergenomic material is specially named, especially repetitive DNA. The main classes are the short (a few hundred nucleotides) interspersed elements (SINES), long (a few thousand nucleotides) interspersed elements (LINES) and tandem (i.e. contiguous) repeats (minisatellites and microsatellites,[7] variable length tandem repeats (VNTR), etc.). These features can be highly specific for individual organisms. Several diseases are associated with abnormalities in the pattern of repeats; for example patients suffering from X syndrome have hundreds or thousands of repeated CGG triplets at a locus (i.e. place on the genome) where healthy individuals have about thirty. The rôle of repetition in DNA is still rather mysterious. One can amuse oneself by creating sentences such as "how did the perch perch?" or "will the wind wind round the tower?" or "this exon's exon was mistranslated",[8] to show that repetition is not necessarily nonsense. The genome of the fruit fly *Drosophila virilis* has millions of repeats of three satellites, ACAAACT, ATAAACT and ACAAATT (reading from the 5′ to the 3′ end), amounting to about 10^8 base pairs, i.e. comparable in length to the entire genome, which does not exceed 2×10^8 base pairs. Another kind of repetition occurs as the duplication, or further multiplication, of whole genes. The apparently superfluous copies tend to acquire mutations vitiating their ability to be translated into a functional protein, wherupon

---

[7]So called because their abnormal base composition, usually greatly enriched in C-G pairs, results in satellite bands near the main DNA bands appearing when DNA is separated on a CsCl density gradient.

[8]Most dictionaries give only one meaning for exon, viz. one of four officers acting as commanders of the Yeomen of the Guard.

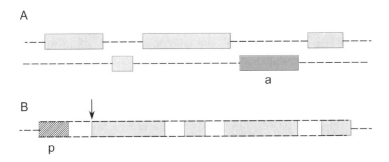

Figure 9.3: Schematic diagram of eukaryotic gene structure. A, the antiparallel double helix. Thick lines (rectangles) represent genes, thin lines intergenomic sequences. B, an expansion of a in A. The shaded rectangles correspond to DNA segments translated to RNA, spliced and translated continuously into proteins. p is a promoter sequence. In reality this is usually more complex than a single nucleotide segment; it may comprise a sequence to which an activator protein can bind (the promoter site proper), but also, more distant ("upstream") from the gene itself, one or more enhancer sites to which additional transcription factors (TF) may bind. All these sites together are called the transcription factor binding site (TFBS). There may be some DNA of indeterminate purpose between p and the transcription start site (TSS), marked with an arrow. Either several individual proteins bind to the various receptor sites, and are only effective all together, or the proteins preassociate and bind *en bloc* to the TFBS. In both cases one anticipates that the conformational flexibility of the DNA is of great importance in determining the affinity of the binding.

they are called pseudogenes.

The satellite sequences of repetitive DNA can occupy an enormous proportion of the genome. In the human, they constitute about 5%; in the horse, about 45%.

Telomere sequences are further examples of repetitive DNA (in humans TTAGGG is repeated for 3–20 kilobases). Between the telomere and the remainder of the chromosome there are 100–300 kilobases of telomere-associated repeats.

The base composition of DNA is very heterogeneous,[9] which makes stochastic modelling of the sequence, e.g. as a Markov chain, very problematical. Certain sequences display extraordinarily long range (10^4 base pairs or more) correlations.[10] This patchiness or blockiness is presumed to arise from the processes taking place when DNA is replicated in mitosis and meiosis (q.v.). It has turned out to be very useful for characterizing variations between individual human genomes. Much of the human genome is constituted from 'haplotype blocks', regions of about 10^4–10^5 nucleotides in which a few (< 10; the average number is 5.5) sequence variants account for nearly all the variation in the world human population. The haplotype map is simply a list of the variants for each block.

Haplotypes are essentially long stretches of DNA characterized by a small number of single nucleotide polymorphisms (SNPs, pronounced 'snips'), i.e. mutated nucleotides. There is an average of about 1 SNP per thousand base pairs in the human genome, hence if they were uncorrelated, in a typical 50 000 base pair haplotype block there would be about 2^{50} (or 4^{50}, depending on whether we are interested in what the base is mutated to) variants—far more variation than is actually found. Hence the pattern of SNPs shows extremely strong constraint—i.e. the occurrences of individual SNPs are strongly correlated with each other. There is considerable interest in trying to correlate haplotype variants with disease, or propensity to disease.

One notes that as much as 98% of the human genome may be identical with that of the ape; one could equally well state that there is more genetic difference between man and woman than between man and ape. To actually derive the vast phenotypic differences between the two from their genomes appears to be as vain a hope as solving the Schrödinger equation for even a single gene.

As an information-bearing sequence, the genome is unusual in that it can operate on itself. The most striking example is furnished by retrotransposons (i.e. transposable elements, whose existence was proposed by McClintock in

---

[9]e.g. Karlin & Brendel.
[10]e.g. Voss.

1948). These gene segments *inter alia* encode a reverse transcriptase enzyme, which facilitates the making of a DNA copy of the sequence. The duplicate sequence is then inserted into the genome; the point of insertion may be remote from that of the gene from which the copy was made. The basis for McClintock's proposal was her observation of rapid variation of the colours of maize kernels from one generation to another; the interpretation of these changes was that a gene coding for colour could be inactivated if a transposon were inserted within it, but the transposon could, with equal facility, be removed during the next round of meiosis, resulting in the reappearance of the colour.

Eukaryotic DNA is organized into chromatin, a protein-DNA complex. The fundamental unit of chromatin structure is the nucleosome, a spheroidal complex made up from eight proteins called histones, around which a stretch of 140–200 base pairs is wrapped. The chromosome is constituted from successive nucleosomes. The protein core of the nucleosome plays a highly significant rôle in the regulation of transcription (§9.6.1). It appears that the histones are precisely positioned relative to the DNA according to its sequence.[11] Further condensation of chromatin occurs in association with long sequences of repetitive DNA to form so-called heterochromatin.

DNA is subject to oxidation, hydrolysis, alkylation, strand breaks etc., countered by various repair mechanisms as discussed below. Molecular machinery (called "SOS") is available to allow replication to proceed despite lesions. Mistakes are the origin of the genotypic mutations leading to the phenotypic variety required by Darwin's theory (see §9.8).

9.3 Molecular mechanisms

In this section DNA replication and recombination will be examined from the molecular viewpoint.

9.3.1 Replication

The molecular mechanism of DNA replication is summarized in table 9.3. Some typical errors—leading to single point mutations—that can occur are summarized in table 9.4.

---

[11]See Audit et al.

Name	operand	operation	operator	result
Premelting	double helix	facilitation	topoisomerase	strand separation
Melting	double helix	facilitation	helicase	strand separation
Synthesis	single strand	nucleotide addition	polymerase	semiconservatively replicated double helix

Table 9.3: DNA replication. Two DNA polymerases are simultaneously active. They catalyse template directed growth in the $5' \rightarrow 3'$ direction. The leading strand is synthesized continuously from $5' \rightarrow 3'$ using the strand beginning with the $3'$ end as the template, whereas the lagging strand is synthesized in short ('Okazaki') fragments using the strand beginning with the $5'$ end as the template. A DNA primase produces a very short RNA primer at the $5'$ end of each Okazaki fragment onto which the polymerase adds nucleotides. The RNA is then removed by an RNAaseH enzyme. A DNA ligase links the Okazaki fragments. A set of initiator proteins is also required to begin replication at the origin of replication. This is of course a simplification. For example, it is estimated that almost 100 (out of a total of ca 6000) genes in yeast are used for DNA replication, and another 50 for recombination.

Name	before	after
Deletion	ABCDEFGH	ABEFGH
Insertion	ABCDEFGH	ABCJFKDEFGH
Inversion	ABCDEFGH	ABCFEDGH
Transposition	ABCDEFGH	ADEFBCGH
Tandem duplications	ABCDEFGH	ABCBCBCDEFGGGGGH

Table 9.4: Some types of chromosome rearrangements with examples. Each letter represents a block of one or more base pairs.

9.3.2 Proofreading and repair

Many proteins are involved in the repair of mismatched and breaks in DNA. Repair takes place after replication, but before transcription. As with Hamming's error-correcting codes, the DNA repair proteins must first recognize the error, and then repair it. It is of primordial importance that DNA is organized into a double helix; the antiparallel strand can be used to check and template repair of mistakes recognized in the other one.

Instead of repair, apoptosis (death of a single cell, as opposed to necrosis, death of many cells in a tissue) of the affected cell may occur.

Concomitant with the work of the specific error recognition and repair enzymes, the entire cell cycle may need to be slowed to ensure that there is time for the repair work to be carried out.

The repair systems are also used to repair damage caused by external factors, e.g. cosmic ray impact, oxidative stress, etc.

The repair mechanisms are essentially directed towards repairing single site errors; there is no special apparatus for eliminating gene duplications, etc. On the other hand, it is not only base mismatches that need to be repaired. Alkylation (methylation) damage could highly adversely affect gene expression and there are enzyme systems (oxidative demethylases and others) for repairing it

Just as certain sequences are more prone to error than others, so are certain erroneous sequences more easily repaired than others. Whereas the quality of a telephone line is independent of the actual words being said, *the fidelity of DNA replication appears to be sequence dependent.* This is used by the genome to explore (via mutations) neighbouring genomes. Hence bioinformatics (applied to genomics) needs a higher-level theory than that provided by existing information theory. A deep, long-range task of bioinformatics is to determine how biological genomes are chosen, such that they are suited to their tasks.

Unreliable DNA polymerase is a distinct advantage for producing new antibodies (somatic hypermutation), and for viruses needing to mutate rapidly in order to evade host defences. In a soup of self-replicating molecules, Eigen has shown that there is a error rate threshold above which an initially diverse population of molecules cannot converge onto a stable, optimally replicating one (a quasispecies).

Problem. What are the implications of a transcription error rate estimated as 1 in 10^5? (In contrast, the error rate of DNA replication is estimated as 1 in 10^{10}.) Calculate the proportion of proteins containing the wrong amino acids due to mistakes in transcription, assuming that translation is perfect.

Compare the result with a translation error rate estimated as 1 in 3000.

9.3.3 Recombination

Homologous recombination is a key process in genetics whereby the rearrangement of genes can take place. It involves the exchange of genetic material between two sets of parental DNA (during meiosis). The mechanism of recognition and alignment of homologous (i.e. with identical, or almost identical, nucleotide sequences) sections of duplex (double stranded) DNA is far less clear than the recognition between complementary single strands, but may depend on the pattern of electrostatically charged (ionized) phosphates, which itself depends slightly but probably sufficiently on sequence, and can be further modulated by (poly)cations adsorbed on the surface of the duplex.[12]

Following alignment, breakage of the DNA takes place, and the broken ends are then shuffled to produce new combinations of genes. For example, consider a hypothetical replicated pair of chromosomes, with the dominant gene written in majuscule and the recessive allele written in miniscule. If $*$ represents a chromosome break, we have

$$
\begin{array}{ll|l}
1 & ABC \\
2 & ABC \\
3 & abc \\
4 & abc
\end{array}
\rightarrow
\begin{array}{ll|l}
1 & AB*C \\
2 & A*BC \\
3 & a*bc \\
4 & ab*c
\end{array}
\rightarrow
\begin{array}{ll|l}
1 & ABc \\
2 & Abc \\
3 & aBC \\
4 & abC
\end{array}
. \tag{9.1}
$$

There is supposed to be about one crossover per chromosome per meiosis. The stages of recombination are, in more detail:

1. Alignment of two homologous double stranded molecules;

2. Breakage of the strands to be exchanged;

3. Approach of the broken ends to their new partners and formation of a fork (also known as the Holliday junction);

4. Broken ends joined to their new partners;

5. Prolongation of the exchange via displacement of the fork;

6. End of displacement;

---

[12]Kornyshev & Leikin.

7. Breakage of the 3′ extremities;

8. Separation of the two recombinant double strands;

9. Repair of the breaks via reading from the complementary strand.

The process is drawn in figure 9.4.

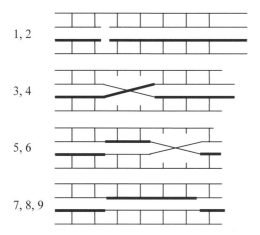

Figure 9.4: Strand exchange in homologous recombination. The numbers refer to the stages described in the text.

Unlike replication, in which occasional single site ('point') mutations occur due to isolated errors, recombination results in changes in large blocks of nucleotides. Correlations between mutations greatly depends on the number of chromosomes. In species with few chromosomes, reshuffling is combinatorially limited and mutations in different genes are likely to be transmitted together from one generation to another, whereas in species with large numbers of chromosomes randomization is more effective. There are also mechanisms whereby chromosome fission and fusion can occur.

9.4 Summary of sources of genome variation

Single site mutations, common to all life forms, may be due to mistakes in duplication (possibly caused by damage to the template base, e.g. due to ionizing radiation). A point mutation is a change in a single base (pair). Note that single insertions or deletions will change the 'reading frame', i.e. all subsequent triplets will be mistranslated.

Microchromosomal and macrochromosomal rearrangements refer to the large-scale changes involving many blocks of nucleotides. Tandem gene duplications may arise during DNA replication, but otherwise the main source for chromosome rearrangement is meiosis.

Prokaryotes mostly do not reproduce sexually, and hence do not undergo meiosis, but on the other hand they are rather susceptible to 'horizontal transfer', i.e. the acquisition of genetic material from other bacteria, viruses etc.[13]

The question of bias in single site mutations is one of great relevance to evolution. The null hypothesis is that any mutation will occur with equal probability. If the mutation is functionally deleterious, according to the Darwinian principle it will not be fixed in the population, and the converse is true for functionally advantageous mutations. Kimura's 'neutral' theory of evolution asserts that functionally neutral (i.e. neither advantageous nor deleterious) mutations will also become incorporated into the genome ('genetic drift').

A similar, but even more intriguing, question can be posed regarding bias in sites of chromosome breakage and crossover. At present, although it is recognized that the likelihood of DNA duplication or moving is sequence dependent, there is no overall understanding of the dependence.

9.5   Gene expression

Gene expression refers to the processes (figure 9.1, d, e, f) whereby proteins are produced ("expressed") from a DNA template. It thus constitutes the bridge between genotype and phenotype. Whenever cells are not preparing for division (and many highly differentiated cells never divide) they are simply living, which means in formal terms that they are engaged in maintaining their essential variables within the domain corresponding to 'alive'. In certain environments, such as ocean floor sediments several kilometres thick, metabolic activity may be barely detectable (many orders of magnitude less than those of familiar laboratory bacteria, or those living parasitically inside a warm-blooded creature). Such environments are moreover unchanging, or barely changing, hence the vital processes could be maintained with very little need to change any of the parameters controlling it.

Most natural habitats show far more variety of conditions however (commonly encountered environmental disturbances include the fluctuating presence of toxic molecules and changes of temperature). Hence, cells need the ability to adapt, i.e. to modify their phenotypes to maintain their essential

---

[13]See Arber.

variables within the vital range. The formal framework for understanding this process was introduced in Chapter 8. Here we examine the molecular mechanisms of regulation that enable adaptation, the control of expression of different proteins as the cell proceeds round its cycle (figure 9.2), and as an organism develops (§9.7): development is a consequence of differential gene expression. The mechanism is essentially the same in all these cases. The entire process of gene expression is facilitated by many enzymes.

9.6 Transcription

The essence of transcription is that RNA polymerases (RNAp) bind to certain initiation sites (sequences of DNA to which their affinity is superior) and synthesize RNA complementary to DNA,[14] taking RNA monomers (nucleotide pyrophosphates) from the surrounding cytoplasm. The enzyme catalyses the formation of a covalent bond between the nucleotide part of the monomer and the extant uncompleted RNA strand, and releases the pyrophosphate part into the cytoplasm as a free molecule. Presumably appropriate hydrogen bonds are formed to the DNA, RNA and incoming nucleotide pyrophosphate, such that if the incoming nucleotide is correctly base-paired with the DNA template, it is held in the correct conformation for making a covalent bond to the extant RNA. The catalysis is reversible but is normally driven in the direction of RNA extension by a constant supply of monomers and the continual removal of the pyrophosphate.

Inition and termination of RNA synthesis are encoded within the DNA sequence. The RNA polymerase is therefore similar in its action to the DNA polymerase in DNA replication. RNAp is a large molecule (M_r almost 500 000).

The RNA folds up as it is synthesized (cf. figure 10.4) but extant structure may have to be disassembled as synthesis proceeds in order to achieve the final structure of the complete sequence.

9.6.1 Regulation of transcription

The key factor in transcriptional regulation is the affinity of RNA polymerase (RNAp) for DNA. The prequisite for RNA production is the binding of RNAp in the initiation zone of the DNA. The binding affinity is *inter alia* influenced by:

---

[14]The transformation is given by: $\downarrow \begin{matrix} \text{A G C T} \\ \text{U C G A} \end{matrix}$.

1. The binding of molecules to the RNAp;

2. The binding of molecules to the DNA initiation zone.

It is convenient to consider transcriptional regulation in prokaryotes and eukaryotes separately.

9.6.2   Prokaryotic transcriptional regulation

The main problem to be solved in prokaryotes is that different genes need to be active under different external conditions and during successive processes in the cell cycle. The primary control mechanism is via promoter sites situated upstream of that part of the DNA that will ultimately be translated into protein (cf. figure 9.3). For genes that need to be essentially constantly transcribed (the so-called housekeeping genes), i.e. coding for proteins that are constantly required, such as those assembling the RNAp complex, there is no hindrance to RNAp binding to the initiation zone and beginning its work; only in exceptional circumstances might it be necessary to arrest production, whereupon a protein (called a repressor) will bind to a sequence within the initiation zone (often immediately preceding the protein coding sequence) called the promoter, preventing the RNAp from binding to the DNA (Sauvageot's principle). Sometimes the transcription factor is simply the gene product. Conversely, for proteins seldom required, such as an enzyme for detoxifying a rarely encountered environmental hazard, the appropriate RNAp will normally have no affinity for the initiation zone, but should the toxin penetrate the cell, it will trigger the binding of a promoting (rather than inhibiting) transcriptional factor (called an activator) to the promoter site, whereupon the RNAp will be able to bind and can start its work.

Sometimes the translation of several genes is controlled by a single promoter. These structures of genes and promoter are called operons.

9.6.3   Eukaryotic transcriptional regulation

The requirements for gene regulation in eukaryotes are more complex, not least because, in a multicellular organism, as the organism differentiates many genes need to be permanently inactivated. Eukaryotes therefore have much richer possibilities for regulating transcription than prokaryotes. The mechanisms fall into three categories:

1. DNA methylation;

2. Chromatin conformation;

3. Binding of complementary ("antisense") RNA to key sites on the DNA;

4. Promoter sites and transcription factors (activators and repressors) as in prokaryotes;

5. Competition for transcription factors by promoter sites on pseudogenes.

Whereas a single RNA polymerase operates in prokaryotes, there are at least three distinct ones in eukaryotes, accompanied by a host of "general transcription factors", which considerably increases the possible combinations of regulatory agents.

DNA methylation

The enzymatic addition of methyl groups to cytosines prevents the gene from being transcribed. This inactivation can be reversed, but some genes are irreversibly inactivated (e.g. in the course of development), e.g. by destruction of the start site, and others are permanently inactivated. It is not understood how these different degrees of inactivation come about. The interrelationship between histone modification (see below) and DNA methylation may play a rôle.

Methylation is the major epigenetic mechanism. Typically, about 80% of the C-G pairs are methylated in human cells, the actual pattern of methylation being highly specific for the cell type.

Chromatin conformation

Long regarded as passive structural elements (despite the fact that the chromosome was known to undergo striking changes in compaction during mitosis), the histones are now perceived as actively participating in the regulation of gene expression. The essential principle is that the histones can be modified and unmodified by covalently attaching and detaching chemical groups, especially to and from the protein "tails" that protrude from the more compact core of the nucleosome. These result in changes in the protein conformation, affecting the conformation of the DNA associated with the histone, and affecting the affinity and accessibility to RNAp. Acetyl groups have attracted particular attention, but methyl and phosphate groups and even other proteins also appear to be involved. The effect of these modifications is to control whether the associated gene is expressed. The modifications are catalysed by enzymes.

Currently there are several ambiguities in the perception of nucleosome-modified gene expression regulation. For example, either acetylation or deacetylation may be required for enabling transcription; and the modification can be local or global (affecting an entire chromosome). Are the effects of the modifications on the ability of transcription enzymes to bind and function at the DNA dependent on the modification of DNA shape, or rigidity, by the modified histones? There may also be proteins other than histones, and also susceptible to modification, associated with nucleosomes. It is appropriate to consider the nucleus as a highly dynamic object full of proteins reacting with and diffusing to, from and along the DNA.

Complementary RNA binding

For many years, the rôles of RNA were thought to be confined to messenger RNA, transfer RNA and ribosomal RNA; remarkably, the very extensive activity of the so-called "noncoding RNA" transcribed from intergenic regions, exons etc. in regulating gene expression was unsuspected until recently. Currently two classes of this small (about two dozen nucleotides) RNA are recognized: "microRNA" (μRNA or miRNA) and small interfering RNA (siRNA). They appear to originate from their own microgenes, or are formed from RNA hairpins (cf. figure 10.5) resulting from mistranscribed DNA.

These small RNA molecules seem to be as abundant as mRNA, and their basic function is to block transcription by binding to complementary DNA sequences, or translation by binding to complementary RNA sequences.

Promoter sites and transcription factors

The affinity of RNAp to DNA is strongly dependent on the presence or absence of other proteins on the DNA, upstream of the sequence to be transcribed (cf. figure 9.3), and associated with the RNAp. The principle of activation and repression by the binding of transcription factors to promoter sites is essentially as in prokaryotes; in eukaryotes more proteins tend to be involved, allowing very fine tuning of expression.

Some molecules can directly interact with mRNA, altering its conformation and preventing translation into protein. This ability can be used to construct a simple feedback control mechanism, i.e. the mRNA binds to its translated protein equivalent. mRNAs able to act in this way are known as riboswitches.

9.6.4 RNA processing

Post-transcriptional modification, or RNA processing, refers to the process whereby the freshly synthesized RNA is prepared for translation into protein. In prokaryotes, translation often starts while the RNA is still being synthesized; in eukaryotes there is an elaborate sequence of reactions preceding translation. In summary, they are capping, 3'-polyadenylation, splicing, and export. Moreover, the whole process is under molecular surveillance and any erroneously processed RNA is degraded back into monomers.

Splicing is needed due to the introns interspersed in the DNA coding for protein. The initially transcribed RNA is a faithful replica of both introns and exons. This pre-mRNA is then edited and spliced (by the spliceosome, which incorporates catalytic RNA and several dozen proteins). The DNA and the enzymes for transcription and post-transcriptional modification are enclosed in the lipid bilayer-based nuclear envelope, whence the edited RNA is exported (as messenger RNA, mRNA) into the cytoplasm for translation.

The current view of of RNA processing is one of extreme complexity, in which many enzymes and enzyme complexes are involved (which of course have their own transcription factors, mRNAs etc.).

Alternative splicing of pre-mRNA is a powerful way of generating variant proteins from the same stretch of DNA; a majority of eukaryotic genes are probably processed in this way.

9.6.5 Translation

The mature mRNA emerges from the nucleus where it is processed by the ribosome, large ($M_r \sim 3 \times 10^6$ in bacteria; eukaryotic ones are larger) abundant (about 15 000 in an *E. coli* cell) protein-RNA complexes. In eukaryotes, ribosomes are typically associated with the endoplasmic reticulum, an extensive internal membrane of the cell. The overall process comprises initiation (at the start codon), elongation and termination (when the stop codon is reached). Elongation has two phases: in the first (decoding) phase, a codon of the mRNA is matched with its cognate tRNA carrying the corresponding amino acid, which is then added to the growing polypeptide; in the second phase the mRNA and the tRNA are translocated one codon to make room for the next tRNA. As established by Crick et al., the mRNA is decoded sequentially in non-overlapping groups of three nucleotides.[15] A messenger RNA may be used several times before it is degraded.

Some of the synthesized proteins are used internally by the cell, e.g. as enzymes to metabolize food and degrade toxins, and to build up structural

---

[15]See table 3.2 for the nucleic acid to amino acid transformation.

members within the cell, such as lipid membranes and cytoskeletal filaments, and organelles such as the chloroplast. Other proteins are secreted to fulfil extracellular functions such as matrix building (for supporting tissues), and other specialized functions, which become more and more complicated as the organism becomes more and more sophisticated. Another group of proteins modulate transcriptional, translational and enzymatic activities. Many proteins have a dual function as a regulator and as something else—e.g. an enzyme may also be able to modulate transcription, either of its own RNA or that of another protein.

About a third of newly synthesized proteins are immediately degraded by proteasomes because they have recognizable folding errors.

9.7   Ontogeny (development)

A multicellular organism begins life as a zygote, which undergoes a series of divisions. The presence of maternal transcription factors regulates the initial pattern of gene activation. Far richer possibilities ensue once several cells are formed, for they can emit and receive substances which activate or inhibit internal processes (including the ability to emit and receive these substances). The developing embryo becomes, therefore, initially a two-dimensional, and then a three-dimensional cellular automaton.

The word 'evolution' was originally coined to describe the unfolding of form and function from single-celled zygote to multicelled adult organism. Since it happens daily and can be observed in the laboratory, it is far more amenable to detailed scientific study than 'evolution' comprising speciation and extinction over geological time scales.

The notion of 'evolution' as the *unfolding* of parts believed to be *already existent in compact form* had already been formalized in 1764 by Bonnet under the name of 'preformation', and had been given a rather mechanical interpretation, i.e. unfolding of a highly compact homunculus produced the adult form.

Later, the term (evolution) came to be used to signify the epigenetic aspects of development. 'Epigenesis' was the alternative to 'preformation', with the connotation of 'order out of chaos'. Both preformation and epigenesis contained the notion of coded instructions, but in the latter, at the time of its formulation the actual mechanism was conceived rather vaguely, e.g. by suggesting the cooperation of 'inner and outer forces'. Nevertheless, it was firmly rooted in the notion of entelechy. In other words, the emphasis was on the potential for development, not on a deterministic path, which is entirely compatible with the cellular automaton interpretation of development. One might also refer to the interaction of genes with their environment.[16] 'Environment' includes constraints set by the physical chemistry of matter in general. William His clearly perceived the importance of general mechanical considerations in constraining morphology.

The term 'ontogeny' was coined by Ernst Haeckel to signify the developmental history of an individual, as opposed to 'phyologeny', signifying the evolution of a type of animal or plant, i.e. the developmental history of an abstract, genealogical individual.

It has been an important guiding principle that ontogeny is a synopsis of phylogeny. Very extensive observations of developing embryos in the eighteenth and nineteenth centuries led to a number of important empirical generalizations, such as von Baer's laws of development (e.g. 'special features appear after the general ones'). It was clear that development embodied different categories of processes with different timescales largely uncoupled from each other: simple growing (the isometric increase of size); growing up (allometric increase,[17] especially important in development of the embryo); and growing older (maturation). By adjusting these timescales relative to each other (heterochrony), different forms could be created.

Much debate has centred around *neoteny*—the retention of juvenile features in the adult animal (paedomorphosis)—and *progenesis*—the truncation of ontogeny by precocious sexual maturation. They can be thought of as respectively retardation and acceleration of development. If organ size (y) is plotted against body size (x), and standard shape defined as $(y/x)_C$, retardation implies that this ratio occurs at larger x, and acceleration, at smaller x. Another form of acceleration is 'recapitulation'—previously adult features are pushed into progressively earlier stages of descendent ontogenies. Table 9.5 summarizes ontogenetic paths.

9.7.1 r and K selection

In an ecological void, i.e. a new environment empty of life, at least of the types we are considering, or a highly fluctuating environment, growth is limited only by the coefficient r in equation 7.5 (r-selection). This circumstance favours progenesis: rapid proliferation at the cost of sophistication, and slight

---

[16]This is a very basic notion that crops up throughout biology. At present there is no satisfactory universal formulation, however, but many interesting models have been proposed and characterized, including those of Érdi and Barna for neurogenesis, and Luthi et al. for neurogenesis in *Drosophila*. All these models of course reduce to the basic formulation for the regulator (§8.2), discussed by Ashby.

[17]Allometric relations are of the type $y = bx^a$, where a and b are constants. $a = 1$ is isometry.

rate				
soma	gonads	effect	morph. result	name
fast	–	acceleration	recapitulation	acceleration
–	fast	truncation	paedomorphosis	progenesis
slow	–	retardation	paedomorphosis	neoteny
–	slow	prolongation	recapitulation	hypermorphosis

Table 9.5: Summary of ontogenetic paths (see text).

acceleration of development leads to a disproportionately greater increase in fecundity.

In an older, more complex ecosystem (with a high density of organisms and intense competition for resources), or a very stable environment, growth is limited by its carrying capacity, i.e. the coefficient K in equation (7.5) (K-selection). This circumstance favours neoteny. Development is stretched out to enable the development of more sophisticated forms. There is no pressure to be fecund; the young offspring would have a very low fitness relative to other species. The most successful beings are likely to be old and wise. The K-selective régime is the scenario for classical progressive evolution, characterized by a primary rôle for increasingly specialized morphology in adaptation, a tendency for size to increase, and hypermorphosis (the phyletic extension of ontogeny beyond its ancestral termination) enabled by delayed maturation.

Both r- and K-selection lead to diminished flexibility: in progenesis, by structural simplification caused by the loss of adult genes; and in the latter, by overspecialization.

A single species in a new, pristine environment simply proliferates until that niche is filled (r-selection). It also explores neighbouring genomes, and if these allow it to more successfully exploit some part of the environment, e.g. at the periphery of the zone colonized, a new species may result. Each new species makes the environment more complex, creating new niches for yet more species, and the environment is thereby transformed into a régime of K-selection.

9.7.2 Homeotic genes

Homeotic genes regulate homeotic transformations, i.e. they are involved in specifying body structures in organisms, homeosis (homoeosis) being a shift in structural development. Homeotic genes encode a protein domain, the homeodomain, which binds to DNA and regulates mRNA synthesis, i.e. it is a transcription factor. The part of the gene encoding the homeodomain is known as the homeobox, or *Hox* gene (in vertebrates). It is a highly conserved motif about 180 bases long. *Hox* and *Hox*-like genes (in invertebrates) are arranged consecutively along the genome and this order is projected onto e.g. the consecutive arrangement of body segments in an insect.

9.8 Phylogeny and evolution

Classical Darwinian theory is founded on two observed facts:

1. There is (inheritable) variety among organisms; and

2. Despite fecundity, populations remain roughly constant;

from which Darwin inferred that population pressure leads to the elimination of descendants less able to survive than slightly different congeners. Formally, therefore, evolution is a problem of selection. Only certain individuals (or species etc.) are selected to survive. It is practically synonymous with 'natural selection', the 'natural' being somewhat redundant.

Modern evolutionary theory is especially concerned with:

1. The levels at which change occurs (e.g. genes, cell lineages, individual organisms, species). Darwin dealt with individual organisms (microevolution); macroevolution deals with mass extinctions;

2. The mechanisms of change, corresponding to the levels. The root of inheritable variation lies in the genes, of course; investigations of mechanisms operating at the higher levels subsume the lower-level mechanisms. The investigation of macroevolution has to deal with unusual events, such as the collision of the earth with a large meteor, and with avalanches of extinctions facilitated by trophic etc. interactions between species;

3. The range of effects wrought by natural selection, and the timescales of change.

Critiques of classical Darwinism are legion. *Inter alia*, one may note that: the selectionist explanation is always a construction *a posteriori*; evidence cited in favour of natural selection is often inconsistent; hence rules are difficult to discern (examples: what is the selectionist advantage of the onerous migration of *Comacchio* eels to the Sargasso Sea for breeding? Why does the cow have multiple stomachs, whereas the horse (a vegetarian of comparable size) only one? Why do some insects adopt marvellous mimicries allowing them to be concealed like a leaf etc., whereas others, such as the cabbage white butterfly, are both conspicuous and abundant? All we can say is that every surviving form must have been viable (i.e. of some selective advantage) or it would not have survived. This is, of course, no proof that it is a product of selection); there appears to be no essential adaptive difference between specialization and nonspecialization—both are found in abundance; selection presupposes all the other attributes of life, such as self-maintenance, adaptability, reproduction etc., hence it is illogical to assert that these attributes are the result of selection; there is no evidence that progression from simple to complex organisms is correlated with better adaptation, selective advantage, or production of more numerous offspring—adaptation is clearly possible at any level of organization, as evinced by the robust survival of very simple forms.

Although the classical theory ascribes competition between peers as a primordial motor of change, decisive evolutionary steps seem to have occurred when the relevant ecological niches were relatively empty, rather than in a period of intense competition.

Arguments of this nature imply that the 'orthodox' view of evolution does not offer a satisfactory explanation of the observed facts. At present we do not have one. It looks likely that principles of self-organization, rooted in the same physico-chemical laws governing the inanimate world, are involved. It would appear to be especially fruitful to focus on the constraints, on which a start has been made by Stephen Jay Gould with his picturesque image of spandrels in vaulted rooms. In well-known buildings such as the San Marco cathedral in Venice their decoration is a notable feature, and contributes so significantly to the overall aesthetic effect that one's first impression is that they were designed into the structure by the architect. They are, however, an inevitable consequence of the vaulting, and were used opportunistically for the decoration, much as feathers, developed to provide thermal insulation, seem to have been used opportunistically for flight, i.e. flight was an exaptation, not an adaptation. Other examples are now known at the molecular level, where existing enzymes are adapted to catalyse new unrelated reactions.

The 'synthetic' theory of evolution (sometimes called gradualism) asserts that speciation is a consequence of adaptation. Species are supposed to arise through the cumulative effects of natural selection acting on a background noise of myriads of micromutations. The genetic changes are not random (in contrast to classical natural selection), nor are they directed towards any goal. Change is opportunistic, i.e. the most viable variants (in a given

context) are selected. Selection takes place in vast populations. The sole
mechanism is intraspecies microevolution.

The synthetic theory is not in accord with the facts of palaeontology.
Ruzhnetsev has emphasized that change is concentrated in speciation events.
The time needed for a new species to become isolated seems to be negligible in
paleontological (let alone geological) time: a few hundred years. Transitional
forms are not observed (on the other hand, certain species have been stable for
more than one hundred million years). Speciation precedes adaptation. This
theory is now usually called 'punctuated equilibrium' (figure 9.5). This is in
sharp contrast to gradualism, which predicts that the rate of evolution (i.e.
the rate of speciation) is inversely proportional to generation time. There is
little evidence for such a correlation, however. On the contrary, for example,
the average species duration $\bar{D}$ for mammals is about 2 My.[18] Their initial
Cenozoic divergence took place over about 12 My, but this would only allow
time for about six speciations, whereas about twenty new orders, including
bats and whales, appeared. Punctuated equilibrium interprets this as the
rapid occupation (by speciation) of niches vacated by dinosaurs in the great
mass extinction at the end of the Cretaceous era.

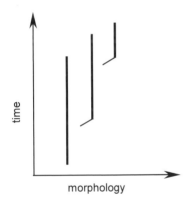

Figure 9.5: Sketch of speciation according to the punctuated equilibrium
concept.

9.8.1   Models of evolution

Typical approaches assume a constant population of M individuals, each of
whose inheritable characteristics are encoded in a string (the genome **s**) of

---

[18]See Stanley for a full discussion.

N symbols. N is fixed, and environmental conditions are supposed fixed too. All the individuals at generation t are replaced by their offspring at generation $t+1$. The state of the population can be described by specifying the genomes of all individuals. Typically, values of M and N are chosen such that the occupancy numbers of most possible genomes are negligibly small. For example, if $N \sim 10^6$ and $M \sim 10^9$, $M \ll 2^N$, the number of possible genomes assuming binary symbols. In classical genetics, attention is focused on a few characteristic traits governed by a few alleles, each of which will be carried by a large number of individuals, and each of which acts independently of the others (hence 'bean bag genetics'). Modelling is able to take much better account of the epistatic interactions between different portions of the genome (which surely corresponds better to reality).

The model proceeds in three stages:

Reproduction: each individual produces a certain number of offspring; the individual α at generation t is the offspring of an individual (the parent) that was living at generation $t-1$, and which is chosen at random among the M individuals of the population;

Mutation: each symbol is modified (flipped in the case of a binary code) at a rate μ; the rate is constant throughout each genome, and the same from generation to generation;

Selection: the genome is evaluated to determine its fitness $W(\mathbf{s}) = e^{F\mathbf{s}/C}$,[19] which in turn determines the number of offspring. C is the selective temperature.

The topography of a fitness landscape is obtained by associating a "height" $F(\mathbf{s})$ with each point $\mathbf{s}$ in genotype space. Various fitness landscapes have been studied in the literature; limiting cases are those lacking epistatic interations (i.e. interactions between genes), and with very strong epistatic interations (one genotype has the highest fitness, the others are all the same). In the latter case the population may form a quasispecies (the term is due to Eigen), consisting of close but not identical genomes. Distances between genomes are conveniently given by the Hamming distance:

$$d_{\mathrm{H}}(\mathbf{s}, \mathbf{s}') = \sum_{i=1}^{N} \frac{(s_i - s_i')^2}{4} , \tag{9.2}$$

---

[19]The fitness of a phenotypic trait is defined as a quantity proportional to the average number of offspring produced by an individual with that trait, in an existing population. In the model the fitness of a genotype $\mathbf{s}$ is proportional to the average number of offspring of an individual possessing that genotype.

and the overlap between two genomes $\mathbf{s}$ and $\mathbf{s}'$ by the related parameter

$$\omega(\mathbf{s}, \mathbf{s}') = \frac{1}{N} \sum_{i=1}^{N} s_i s_i' = 1 - \frac{2d_{\mathrm{H}}(\mathbf{s}, \mathbf{s}')}{N} . \qquad (9.3)$$

ω is an order parameter analogous to magnetization in a ferromagnet. If the mutation rate is higher than an error threshold than the population is distributed uniformly over the whole genotype space ("wandering" régime) and the average overlap $\sim 1/N$; below the threshold, the population lies a finite distance away from the fittest genotype and $\omega \sim 1 - \mathcal{O}(1/N)$.[20] Intermediate between these two cases (none and maximal epistatic interactions) are the rugged landcapes studied by Kauffman.[21] More realistic models need to include changing fitness landscapes, resulting from interactions between species—competition (one species inhibits the increase of another), exploitation (A inhibits B but B stimulates A) or mutualism (one species stimulates the increase of another)—i.e. coevolution.

As presented, the models deal with asexual reproduction. Sex introduces complications, but can in principle be handled within the general framework.

These models concern microevolution (the evolving units are individuals); if the evolving units are species or larger units such as families, then one may speak of macroevolution. There has been particular interest in modelling mass extinctions, which may follow a power law (the number n of extinguished families $\sim n^\gamma$, with γ equal to about -2 according to current estimates. Bak & Sneppen invented a model for the macroevolution of biological units (such as species) in which each unit is assigned a fitness F, defined as the barrier height for mutation into another unit. At each iteration, the species with the lowest barrier is mutated, i.e. is assigned a new fitness, chosen at random from a finite range of values. The mean fitness of the ecosystem rises inexorably to the maximum value, but if the species interact, and a number of neighbours are also mutated (regardless of their fitnesses (this simulates the effect of, say, the extinction of a certain species of grass on the animals feeding exclusively on that grass) the ecosystem evolves such that almost all species have fitnesses above a critical threshold, i.e. the model shows self-organized criticality. Avalanches of mutations can be identified and their size follows a power law distribution, albeit with $\gamma \sim -1$. Hence there have been various attempts to modify the model to bring the value of the exponent closer to the value (-2) believed to be characteristic of the earth's prehistory.

---

[20] See Peliti for a comprehensive treatment.
[21] See §7.2. See Jongeling for a critique.

9.8.2   Sources of genome variation

Non-Darwinian evolution ascribes the major rôle in molecular evolution to "genetic drift"—random ("neutral") changes in allele frequency. Classically, it is questionable whether genotypic differences without an effect on phenotype can affect fitness (in any sense relevant to evolution). One should bear in mind that one of the engines of evolution, natural selection, operates on phenotype not genotype (to a first approximation at least), and therefore genes on their own are only the beginning of understanding life: it is essential to understand how those genes are transformed into phenotype. To survive, however, a species or population needs adaptedness (to present conditions), (genetic) stability, and (the potential for) variability. Without stability, reproductive success would be compromised. Genetic variability is of course antithetical to stability, but phenotypic variability, reflecting control over which portion of the protein repertoire will be expressed, determines the range of environments in which the individual can survive, and hence is equivalent to adaptedness to future conditions. The eukaryotic genome, with its resources of duplicate genes, pseudogenes, transposable elements, exon shuffling, polyploidy etc. possesses the potential of phenotypic variability while retaining genetic stability. Prokaryotes lack these features, but on the other hand can readily acquire new genetic material from their peers or from viruses.[22]

9.8.3   The origin of proteins

The random origin hypothesis[23] asserts that proteins originated by stochastic processes according to simple rules, i.e. that the earliest proteins were random heteropolymer sequences. This implies that their length distribution is a smoothly decaying function of length (determined by the probability that a stop codon will occur after a start codon has been encountered, in the case of templated synthesis without exons). On the other hand, the probability that a sequence can fold into a stable globular structure is a slowly increasing function of length up to about 200 amino acids, after which it remains roughly constant. Convolution of these two distributions results in a length distribution remarkably similar to those of extant proteins.

---

[22]Discussed by Arber.
[23]See White.

9.9 Geological eras and taxonomy

In this section are appended tables of the major groupings of living and growing things and the geological eras of the Earth.

Three lineages are recognized, the archaea (represented by extremophilic prokaryotes, formerly known as archaebacteria), the eubacteria ("true" bacteria, to which the mitochondria and chloroplasts are provisionally attributed) and the eukaryotes (possessing true nuclei). The eukaryotic kingdoms are animalia (metazoa), plantae, fungi and protista (protozoa, single-celled organisms, including algae, diatoms, flagellates, amoebae etc.). The approximate numbers of species of these different kingdoms are currently estimated as 10^7 (metazoa), 2.5×10^5 (plantae), 2×10^5 (protozoa) and 5×10^4 (fungi).

Name	example (1)	example (2)
kingdom	animalia (metazoa)	plantae (green plants)
phylum	chordata	angiospermophyta
subphylum	vertebrata	–
class	mammalia	monocotyledonae
order	primates	asparagales
suborder	anthropoidae	–
superfamily	hominoidae	–
family	hominidae	alliaceae
genus	*Homo*	*Allium*
species	*sapiens*	*sativum*
individuals	Fred Bloggs	–

Table 9.6: The hierarchical scheme of the descriptive taxonomy of eukaryotes. Examples are given for an individual human being and the culinary garlic.

Phylum	characteristic	examples
porifera	no permanent tissue	sponges
coelenterata (cnidaria)	2 or 3 layers of cells	nematode worms
ctenophora	2 or 3 layers of cells	comb jellies
annelida	mesoderm has a cavity	earthworms
arthropoda ($\sim \frac{4}{5}$ of all animal species)	jointed limbs	insects, crustaceans, arachnids
mollusca	true coelom	snails, octopus
echinoderma	urchin-skinned	starfish
chordata	backbone, skull	–

Table 9.7: The major divisions (phyla) of animals. The chordata (craniata) are subdivided into subphyla including the vertebrata, whose classes comprise the familiar agnatha (lampreys etc.) fish, amphibians, reptiles, birds and mammals.

name	epoch[a]	events	new or dominant life forms
–	4300	Earth formed	none
Phanerozoic			
–	3500?	–	first life
–	3000?	–	stromatolites
–	2500?	–	mitochondria
–	2000?	–	bacteria
Palaeozoic			
Cambrian	570–500	–	trilobites
Ordovician	500–440	–	–
Silurian	440–410	–	fish, land plants
Devonian	410–345	–	–
Carboniferous	345–280	abundant plants	giant insects, reptiles
Permian	280–225	Panguea[b], hot & dry	reptiles dominant
Mesozoic			
Triassic	225–190	Gondwanaland[c]	–
Jurassic	190–134	warm	gymnosperms & ferns
Cretaceous	135–65	mass extinction at end	birds, dinosaurs
Cenozoic (tert.			
Palaeocene	65–54	volcanoes	many
Eocene	54–38	separation of Eurasia	high diversity
Oligocene	38–26	cooling	low diversity
Miocene	26–7	continental collisions	–
Pliocene	7–2.5	Himalayas, Alps	elephants, *Australopicethus*
Cenozoic (quat.)			
Pleistocene	2.5–0.01	last ice age	woolly mammoth
Holocene	0.01–pres.	–	–

Table 9.8: History of the Earth and earthly life.
[a] In millions of years before present. [b] The single supercontinent. [c] The great southern continent.

Chapter 10

The molecules of life

10.1 Molecules and supramolecular structure

Table 10.1 gives some approximate values for the atomic composition of a cell. The atomic composition represents a highly reductionist view, somewhat akin to asserting that the informational content of *Macbeth* is $-\sum_{\text{alphabet}} p_i \log_2 p_i$, where p_i is the normalized frequency of occurrence of the ith letter of the alphabet. The next stage of complexity is to consider molecules (tables 10.2 and 10.3). This is still highly reductionist, however—it corresponds to calculating Shannon entropy from the vocabulary of *Macbeth*. Words are, however, grouped into sentences, which in turn are arranged into paragraphs. The cell is analogously highly structured—molecules are grouped into supramolecular complexes, which in turn are assembled into organelles. This structure, some of which is visible in the optical microscope, but which mostly needs the higher resolution of the electron microscope, is often called ultrastructure. It is difficult to quantify, i.e. assign numerical parameters to it, with which different sets of observations can be compared. The human eye can readily perceive drastic changes in ultrastructure when a cell is subjected to external stress, but generally these changes have to be described in words.

The most prominent intracellular structural feature is the system of lipid bilayer membranes, such as the endoplasmic reticulum. Also prominent are the proteins such as actin, which form large filamentous structures constituting a kind of skeleton (the cytoskeleton). There are also many more or less compact (globular), large multiprotein complexes (e.g. the proteasome). Furthermore, proteins may be associated with lipid membranes, or with the DNA. These structures are rather dynamic, i.e. there is ceaseless assembly and disassembly, depending on the exigencies of survival. Some of them are described in more detail under the descriptions of the individual classes of molecules.

The interior of the cell is an exceedingly crowded milieu (compare the quantities of molecules with the dimensions given in table 10.4). Although water constitutes about 70% of a typical cell, very little of this water is free, bulk material. The very high concentrations of molecules and macromolecules ensure that the cytoplasm is a highly viscous medium. Moreover, most of the macromolecules, e.g. proteins, are attached to larger structures such as the internal membranes. Kempner and Miller's classic experiments, in which they centrifuged intact cells to separate macromolecules from the water, demonstrated this very clearly—hardly any macromolecules were found in the aqueous fraction. This was in sharp contrast to the result of the traditional biochemical procedure of destroying all ultrastructure by mechanical homogenization, yielding an aqueous cytosol containing many dissolved enzymes.

The effect of the ultrastructure is twofold: to divide the cell up into compartments, not hermetically separated from one another but allowing access to different zones to be controlled; and to provide two-dimensional surfaces on which searching for and finding reaction partners is far more efficient, as was seen in Chapter 6.

The separation of the macromolecules, which of course plays a crucial part in experimental bioinformatics, is dealt with in Part III.

Element	Rel. atomic fraction
H	100 000
C	5300
O	1600
N	1300
P	130
K,Na	80
S	40
Fe	5
Cu	1

Table 10.1: Atomic composition (selected elements) of a typical dried microbial cell.

Molecule	wt %	mol %	M_r	no types	no molec.
DNA	1	–	3×10^9	1	1
RNA	6	–	(10^5)	500	250 000
protein	15	–	5×10^4	1000	2×10^6
saccharide	3	–	(10^4)	50	5000
lipid[a]	2	0.1	10^3	40	2×10^7
small[b]	2	1	10^2	500	10^7
water	70	98.9	18	1	2×10^{10}

Table 10.2: Molecular composition of a typical microbial cell. The components are not uniformly dispersed in the cell. Parentheses indicate approximate means of very broad ranges.
[a] including liposaccharides. [b] metabolic intermediates, inorganic ions etc.

Polymer	monomer	variety	length	bond variety[a]
DNA	nucleotide	4	2000	1
RNA	nucleotide	4	2000	1
protein	amino acid	20	200	1
polysaccharide	monosaccharide	~ 10	20	~ 3

Table 10.3: Some characteristics of the macromolecules of a cell. A nucleotide consists of a base, a sugar, and one or more phosphate groups. The variety resides solely in the bases. An amino acid consists of a backbone part, identical for all except proline, and a side chain (residue) in which the variety resides.
[a] i.e. the type of bonding between monomers.

Property	Mole fraction
Shape	sphere
Density	1.025 g/cm^3
Radius	5 μm
Volume	5×10^{-16} m^3
Surface charge	-10 fC/μm^2
Coat material	polysaccharide
Coat thickness	10 nm
Coat charge density	-5 MC/m^3

Table 10.4: Morphology etc. of a typical eukaryotic cell. A typical prokaryote, such as the organism specified in Table 10.2 would have a diameter about ten times smaller.

10.2 Water

As seen from table 10.2, water is overwhelmingly dominant in the cell. Water (H_2O) is a very unusual substance, as can be inferred from its unusually high boiling point (compared with other molecules of comparable size) and large specific heat. Salient features are the great polarity of the molecule—the bond between the oxygen and the hydrogen has a very strong ionic character. The electrostatic attraction between the positively charged hydrogen ($\delta+$) and the negatively charged electron lone pair on the oxygen ($\delta-$) constitutes the hydrogen bond (figure 10.1). It can be thought of as a redistribution of electron density from the covalent O–H bond to the zone between the H and the neighbouring O. The loss of electron density from the covalent O–H bond results in a weaker, more slowly vibrating bond.

Each water molecule can simultaneously accept and donate two hydrogen bonds (each hydrogen is a donor, and the oxygen bears two lone electron pairs). In flawless ice, the water molecules are H-bonded together in a tetrahedral arrangement.

The O–H infrared spectrum (of HOD in liquid D_2O) gives a very broad distribution of energies, implying a continuum from ice-like to nonbonding. In pure water at room temperature, about 10% of the O–H groups and lone pairs (LP) are nonbonded; close to the boiling point this percentage rises to about 40.

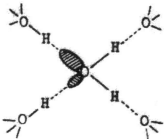

Figure 10.1: A water molecule hydrogen bonded to its congeners. The hydrogen atom is typically 0.10 nm from the oxygen to which it is covalently bonded, and 0.18 nm from the neighbouring oxygen to which it is hydrogen bonded. The energy of the hydrogen bond is about 0.1 eV, i.e. about $4k_BT$ at room temperature, or about 2.4 kJ/mol (after Symons).

Bonded and nonbonded ions are in equilibrium:

$$H_2O_{\text{fully bonded}} \rightleftharpoons OH_{\text{free}} + LP_{\text{free}} \qquad (10.1)$$

where subscript free denotes nonbonded. LP_{free} and OH_{free} are respectively an electron donor (Lewis base) and electron acceptor (Lewis acid) and hence can interact with other species present in solution. An ion pair such as KCl interacts with both LP_{free} and OH_{free} in roughly equal measure, hence KCl does not perturb the equilibrium 10.1, whereas (to take an extreme case) $NaB(C_6H_5)_4$ can only interact with LP_{free}, hence increasing the concentration of free OH groups. These sort of interactions have profound implications for macromolecular structure, as will be seen (§10.5).

10.3 DNA

Deoxyribonucleic acid is considered to the ultimate repository of potentially meaningful information in the cell. DNA is poly(deoxyribonucleic acid), and the information is conveyed by the particular sequence of bases of the polymer. Each monomer unit has three parts: base, sugar and phosphate (Fig. 10.2). The sugar (deoxyribose) and phosphate are always the same; the possibility of storing information arises through varying the base, for which there are four possibilities, the purines adenine (A) and thymine (T), and the pyrimidines cytosine (C) and guanine (G). The strand running from 5' to 3' is called the 'sense' strand, i.e. it is used to specify protein sequences (via RNA), and the other one the antisense (antiparallel) strand. Mainly

only one strand encodes this information and the complementary one serves
to correct damage.

Each base has the very important property of being able to hydrogen bond
with one of the other three, the complementary base, significantly better than
to any of the others. This is perhaps the purest, most elementary example of
'molecular recognition'. Hence a polymerized chain of monomers can serve
as a template for the assembly of a complementary strand. The purine pairs
are linked by only two hydrogen bonds (H-bonds), whereas the pyrimidines
are linked by three (figure 10.3). This means that the C-G base pairing melts
(i.e. the H-bonds are broken) at a higher temperature than the A-T one.

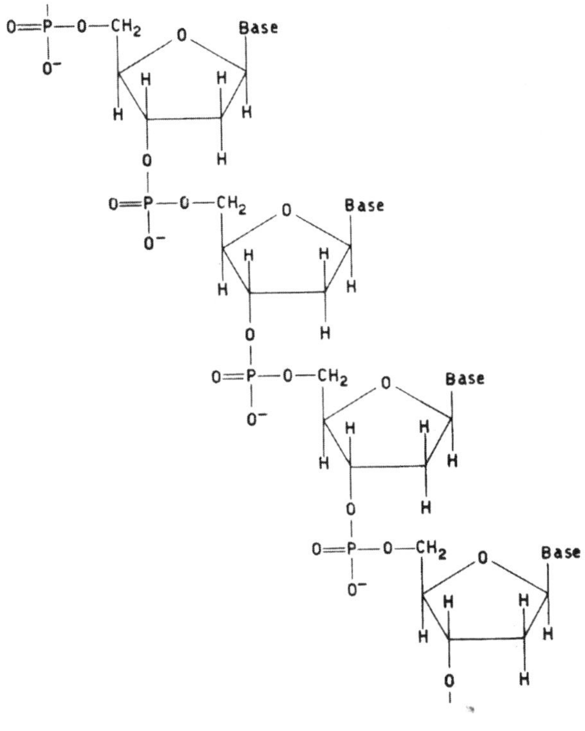

Figure 10.2: Polymerized DNA. The so-called 3′ end is at the upper left end,
and the 5′ at the lower right (after Ageno, 1967).

As expected from their aromatic structure, the bases are planar. Figure
10.4 shows the formation of the double helix. The genes of most organisms
are formed by such a double helix. The melting of the H-bonds as the temperature is raised is highly cooperative (due to the repulsive electrostatic force

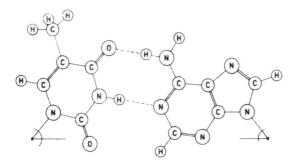

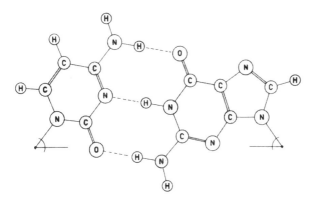

Figure 10.3: The hydrogen-bonding patterns of complementary bases (adenine, guanine, cytosine and thymine moving round clockwise from the upper right). In RNA, uracil replaces thymine (i.e. the methyl group on the base is replaced by hydrogen) and the ribose has a hydroxyl group (after Ageno, 1967).

between the charged phosphate groups). On average, the separation into single stranded DNA occurs at about 80 °C (at about 90 °C for sequences rich in C–G pairs, and at about 65 °C for sequences rich in A–T pairs). These melting temperatures are lower at extremes of pH. Melting leads to complete separation of the two chains, which is made use of in artificial gene manipulation, as discussed in Part III. During *in vivo* replication, as discussed in the previous chapter, the chains are only separated locally.

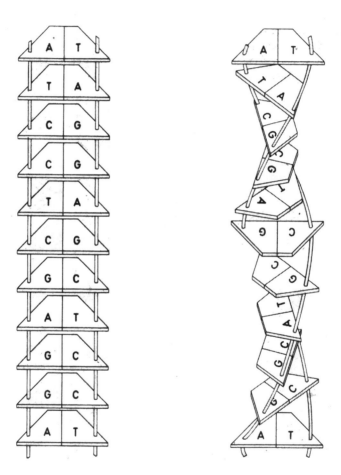

Figure 10.4: A stack of polymerized base pairs (left) distorted (right) by slightly twisting in order to form the double helix (after Ageno, 1967).

Table 10.5 summarizes some significant discoveries relating to DNA.

Event	Year	Principal worker(s)
Nuclei contain an acidic substance	1869	Miescher
A tetranucleotide structure elucidated	1919	Levene
DNA identified as genetic material	1944	Avery
First protein[a] sequenced	1953	Sanger
DNA double helical structure	1953	Watson & Crick
Sequence hypothesis, central dogma	1957	Crick
Semiconservative replication	1958	Meselson & Stahl
DNA polymerase isolated	1959	A. Kornberg
Sequential reading of bases	1961	Crick
First protein sequence data bank	1965	–
Genetic code decrypted	1966	Crick
First entire genome[b] sequenced	1995	–
First multicellular genome[c]	1999	–

Table 10.5: Some milestones in molecular bioinformatics.
[a] Insulin. [b] *Haemophilus influenzae.* [c] *Caenorhabditis elegans.*

10.3.1   Structure of DNA

It is now recognized that the structure, especially the sequence- and mod-
ification-dependent rigidity (bending modulus) plays a profound rôle in the
fidelity of replication, the regulation of transcription, and the movement of
DNA through crowded milieux. The last aspect is of practical importance in
DNA fractionation for sequencing, etc.

Under typical conditions of temperature, acidity, salt concentration etc.
prevailing in cells, the right-handed (Watson and Crick) double helix is the
most stable structure, but others exist, such as the left-handed helix (Z-
DNA). Circular DNA can be supercoiled; differing degrees of supercoiling
affect the accessibility of the sequence to RNA polymerase etc., and is thus a
regulatory feature. There are several enzymes (topoisomerases, gyrases and
helicases) for changing DNA topology.

Double stranded DNA is a rather rigid polymer, yet despite its length if
stretched out in a straight line (about 1.2 mm for the DNA of *E. coli*), it is
nevertheless packed into a cell only about 1 μm long. Human DNA would
be about 1 m long.

A prominent feature of DNA is its high negative charge density due to the
phosphate groups along the backbone. This gives DNA an ionic strength-
dependent rigidity, which is also a factor affecting transcription and translation.

The rigidity can be quantified by the persistence length p, which depends
on Young's modulus E:

$$p = EI_s/(k_BT) \qquad (10.2)$$

where I_s is the moment of inertia ($= \pi r^4/4$ for a cylinder of radius r), k_B is
Boltzmann's constant and T the absolute temperature. For DNA, $r \approx 1.2$
nm and $E \approx 10^6$ N/m, giving $p \approx 60$ nm. The radius of gyration R_g of the
polymer as a gaussian coil is given by $(Lp/6)^{1/2}$.

A mixture of different molecules of DNA is usually separated into its
components using gel electrophoresis, in which the DNA is driven by an electric field through a hydrogel (usually polyacrylamide or agarose). Recently,
model environments have been created from arrays of precisely positioned
microfabricated pillars. Long polymers in such confined media move by reptation (rather like a snake moving through tall stiff grass—it is constrained
laterally but can move along its length), in which they are confined to sliding
along an imaginary tube between the pillars. The diffusivity D is, as usual

$$D = k_BT/\delta \qquad (10.3)$$

where δ is the drag coefficient, equal to $2\pi\eta L$, η being the viscosity of the solvent. The time for the polymer to diffuse out of its tube of length L is

$$\tau = L^2/(2D) , \qquad (10.4)$$

but in that time the polymer would have moved a distance equal to R_g if it had formed a gaussian coil; the effective diffusion coefficient in the gel is then found from $D_{gel}/D = (R_g/L)^2$, hence

$$D_{gel} = \frac{pk_BT}{12\pi\eta L^2} . \qquad (10.5)$$

Under the action of a relatively weak electric field, and provided L is not too great, the mobility of the DNA in the gel is

$$\mu = \frac{\sigma p}{\sqrt{12\pi\eta L}} , \qquad (10.6)$$

where σ is the charge per unit length of the DNA.[1]

10.4   RNA

Ribonucleic acid, RNA, is rather similar to DNA. The most prominent difference is that the sugar is ribose rather than deoxyribose, and that uracil rather than thymine is used as one of the two purine bases. These differences have considerable structural consequences. RNA does not occur as double helices: instead base pairing is internal, forming parallel strands, loops and bulges (Fig. 10.5). It can therefore adopt very varied three dimensional structures. It can pair (hybridize) with DNA.

RNA has five main functions: as a messenger (mRNA), acting as an intermediary in protein synthesis; as an enzyme (ribozymes); as part (about 60% by weight, the rest being protein) of the ribosome (rRNA); as the carrier for transferring amino acids to the growing polypeptide chain synthesized at the ribosome (tRNA); and as a modulator of DNA and mRNA interactions (small interfering RNA, siRNA).

Since ribozymes can catalyse their own cleavage, RNA can form evolving systems and hence it has been suggested that the earliest organisms used RNA rather than DNA as their primary information carrier. Indeed, some extant viruses do use RNA in that way.

---

[1]For polymers confined by their congeners, a given chain can slowly escape from its tube by Brownian motion: the mobility μ of the whole chain N monomers long is μ_1/N, where μ_1 is the mobility of one monomer. Hence from the Einstein relation $D_{tube} = \mu_1 k_B T/N$, and the relaxation time (to which viscosity is proportional) for tube length L ($\sim N$) to be lost and created anew $\tau_{tube} \sim L^2/D = NL^2/(\mu_1 k_B T) \sim N^3$, in contrast to small molecules not undergoing reptation, for which $\tau \sim N$.

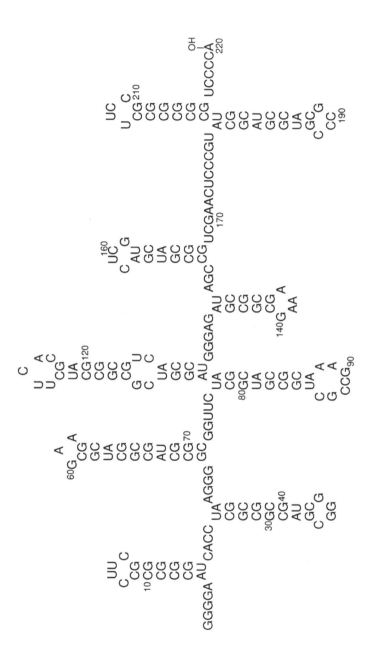

Figure 10.5: A piece of RNA (from the Qβ replicase MDV-1) showing the characteristic loops formed by single strand base pairing.

10.4.1 Folding of RNA

A least action approach, i.e. minimizing the integral of the Lagrangian $\mathcal{L}$ (i.e. the difference between the kinetic and potential energies), has been successfully applied to predicting RNA structure. The key step was finding an appropriate expression for $\mathcal{L}$. The concept can be illustrated by focusing on loop closure, considered to be the most important folding event. The potential energy is the enthalpy, i.e. the number n of contacts (base pairings), and the entropy yields the kinetic parameter. Folding is a succession of events in which at each stage as many new intramolecular contacts as possible are formed, while minimizing the loss of conformational freedom (the principle of sequential minimization of entropy loss, SMEL). The entropy loss associated with loop closure is ΔS_{loop} (and the rate of loop closure $\sim \exp(\Delta S_{\mathrm{loop}})$); the function to be minimized is $\exp(-\Delta S_{\mathrm{loop}}/R)/n$. A quantitative expression for ΔS_{loop} can be found by noting that the N monomers in an unstrained loop ($N \geq 4$) have essentially two possible conformations, pointing either inwards or outwards. For loops smaller than a critical size N_0, the inward ones are in an apolar environment, since the enclosed water no longer has bulk properties, and the outward ones are in polar bulk water; hence the electrostatic charges on the ionized phosphate moieties of the bases will tend to point outwards. For $N < N_0$, $\Delta S_{\mathrm{loop}} = -RN \ln 2$, and for $N > N_0$, the Jacobson-Stockmayer approximation based on excluded volume yields $\Delta S_{\mathrm{loop}} \sim R \ln N$. This allows $\mathcal{L}$ to be completely specified.[2]

10.5 Proteins

Proteins are appropriately named after the mythological being Proteus, who could assume many forms. The functions of proteins are structural, catalytic, etc. The catalytic functions are especially important, for almost all of the other molecules of life, as well as small metabolites, are synthesized with their help. A rough overview of the protein world reveals the existence of:

Small polypeptides typically with no definite structure, acting as hormones, toxins etc. Examples: bradykinin, mellitin;

Globular proteins typically able to assume a small number of stable configurations. This is the most numerous and varied class of proteins, comprising enzymes, transporters, regulators, motors, etc. Examples: glucose oxidase, haemoglobin, tumour necrosis factor α, kinesin;

---

[2]See Fernández & Cendra.

Fibrous proteins which may be very long. They often have modular structures with many identical modules. Their rôle is mostly structural, but they actively interact with e.g. neurites growing on them, i.e. as basement membranes they show chemical specificity. Examples: collagen, laminin;

Membrane proteins which are also globular, but permanently embedded (tranversally) in a lipid bilayer membrane. They mainly function as channels and signal transducers. Examples: bacteriorhodopsin, porin.

In the remainder of this subsection, we shall concentrate on globular protein structure.

10.5.1 Amino acids

The basic structure of an amino acid is H_2N-$C^{(\alpha)}$HR-COOH. At physiological pH it exists as a zwitterion, H_3N^+-$C^{(\alpha)}$HR-COO$^-$. R denotes the variable residue; except for glycine (R=H), the $C^{(\alpha)}$ is asymmetric and hence chiral. The different residues are given in table 10.6

Amino acid polymerization takes place via elimination of water and the formation of the so-called peptide bond. Hence a tripeptide wuth residues R_1, R_2 and R_3 has the structure H_2N–$C^{(\alpha)}$HR$_1$–CO–N–$C^{(\alpha)}$HR$_2$–CO–N–$C^{(\alpha)}$HR$_3$–COOH. Amino acids polymerized into a polypeptide chain are usually called peptides. The CO–N bond is in resonance with the C=O bond and is therefore rigid, the CO–N triatom system being planar, but the N–$C^{(\alpha)}$ and $C^{(\alpha)}$HR$_1$–CO bonds are free to rotate independently. Two dihedral angles, ϕ and ψ respectively, per amino acid therefore suffice to completely characterize the conformation of a polypeptide chain. A Ramachandran plot of *psi* versus ϕ can be constructed for each amino acid showing the allowed conformations; constraints arise due to the overlaps between the atoms attached to the N–$C^{(\alpha)}$–C backbone.

The hydrogen bonding capabilities of the backbone and residues of the amino acids are shown in figure 10.6.

10.5.2 Protein folding and interaction

Proteins are synthesized *vivo* by the consecutive addition of amino acids to form an elongating peptide chain with the conformation of a random coil in the aqueous cytoplasm. Native globular proteins are compact stable structures with no or very few polar residues in their interior. The transition from a random coil to an ordered globule is called folding.

Name	a	b	polarity^c	formula^d
alanine	ala	A	A	$-CH_3$
arginine	arg	R	+	$-(CH_2)_3-NH-C(NH_2)_2^+$
asparagine	asn	N	P	$-CH_2-CONH_2$
aspartic acid	asp	N	−	$-CH_2-COO^-$
cysteine	cys	C	P	$-CH_2-SH$
glutamine	gln	Q	P	$-(CH_2)_2-CONH_2$
glutamic acid	glu	E	−	$-(CH_2)_2-COO^-$
glycine	gly	G	A	$-H$
histidine	his	H	+	$-CH_2-[C_3N_2H_3]^+$
isoleucine	ile	I	A	$-CH(CH_3)-CH_2-CH_3$
leucine	leu	L	A	$-CH_2-CH(CH_3)_2$
lysine	lys	K	+	$-(CH_2)_4-NH_3^+$
methionine	met	M	A	$-(CH_2)_2-S-CH_3$
phenylalanine	phe	F	A	$-CH_2-\phi$
proline	pro	P	A	$-[C_3NH_7]^e$
serine	ser	S	P	$-CH_2-OH$
threonine	thr	T	P	$-CH(OH)-CH_3$
tryptophan	trp	W	A	$-CH_2-[C_8NH_6]$
tyrosine	tyr	Y	P	$-CH_2-\phi-OH$
valine	val	V	A	$-CH(CH_3)_2$

Table 10.6: The amino acids in alphabetical order. ϕ denotes a benzene ring. Square brackets denote a ring structure.
[a] Three letter code. [b] One letter code. [c] A, apolar; P, polar; +, positively charged (at physiological pH); −, negatively charged. [d] Of the side chain. [e] Incorporates the backbone -NH$_2$ in a ring structure.

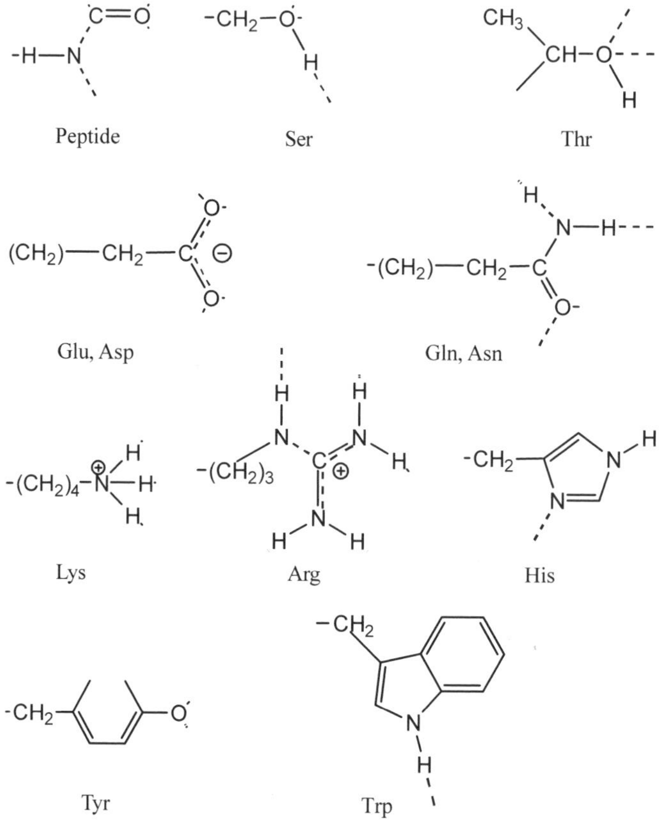

Figure 10.6: Hydrogen bonding capabilities of the peptide backbone and the polar residues (after Baker and Hubbard). Residues not shown are incapable of hydrogen bond formation.

The governing feature of the polypeptide is the ability of the peptide unit –N–C–C(=O)– to accept and donate hydrogen bonds. The ith residue in a chain can bond with the $i \pm 3$th residues to form the α-helix, due to geometrical constraints. This is the primary structural element of proteins. Very simple polypeptides (e.g. polyalanine) form a pure α-helix. Most globular proteins, made up of many different amino acids, contain short α-helices joined by turns—short polypeptide segments of no special structure. The other main structural element is the β-sheet, in which the hydrogen bonds are formed between peptides distant along the chain.[3]

The formation of these hydrogen bonds has to, and does, take place in the presence of water, which is of course present in huge excess. Water is an excellent donor and acceptor of hydrogen bonds and strongly competes for the intraprotein ones. Successful folding therefore depends on the ability of the protein to isolate the structurally important hydrogen bonds from water; in other words, structural integrity requires that the backbone H-bonds are kept dry. The energetic importance of H-bond wrapping (i.e. protection from water) can be seen by noting that the energy of a hydrogen bond is strongly context-dependent. In water, it is about 2 kJ/mol; *in vacuo* it increases eight to tenfold. Wrapping will therefore greatly contribute to the enthalpic stabilization of globular protein conformation.

A poorly desolvated hydrogen bond is called a dehydron.[4] The dehydron is underwrapped, overexposed to water (i.e. wet), because there are insufficient apolar groups in its vicinity. The only way for a protein to diminish the presence of water around a hydrogen

bond is to bring apolar residues unable to form hydrogen bonds with water into its vicinity. Hydrophobic (apolar) groups, such as methyl, ethyl etc., are powerful H-bond enhancers. The *dehydronic force* is thus a three body force involving the hydrogen bond donor, the hydrogen bond acceptor, and the apolar residue. It is formally defined as the drag exerted by a dehydron on a test residue, i.e.

$$F = -\nabla_{\mathbf{R}} \left(\frac{1}{4\pi\epsilon\mathbf{R}} \frac{qq'}{r_0} \right) \tag{10.7}$$

where $\mathbf{R}$ is the position of the test residue (hydrophobic) measured perpendicularly from the H-bond, q and q' are the net charges and r_0 the O-H distance of the H-bond. Typically F is about 7 pN at $\mathbf{R} = 6$ Å.

---

[3]As shown in figure 10.6, some residues can also participate in hydrogen bonding, but the backbone peptide H-bonds (or potential H-bond donors and acceptors) are of course far more numerous, and hence significant.

[4]The dehydron concept is due to Fernández. See for example Fernández & Scott, 2003 and Fernández et al., 2002 and 2003.

The three dimensional structure of a protein (as encoded in a pdb file) can be interrogated to reveal dehydrons. Hydrogen bonds are operationally defined as satisfying the criteria of an N–O distance of 2.5 $\sim$ 3.5 Å, and the angle between the NH and CO bonds equal to 45°. The dehydration domain of a hydrogen bond is defined as two spheres of equal size centred on the $C^{(\alpha)}$s of the amino acids paired by the H-bond. The radius of the spheres (around 6.5–7 Å) is chosen to slightly exceed the typical distance between nonadjacent $C^{(\alpha)}$s, hence the spheres necessarily interact. The extent of wrapping is given by the number ρ of hydrocarbon groups within the dehydration domains. A well-wrapped H-bond has $\rho = 15$; most soluble monomeric globular proteins have a ρ around this value, averaged over all the backbone H-bonds.

Wrapping defects are decisive determinants of protein-protein (and other) interactions. If the stable conformation of a globular protein is such that there are some unavoidably underwrapped hydrogen bonds on its solvent-acessible surface, then that protein will be sticky; the underwrapped hydrogen bonds will be the hotbeds of stickiness.[5] Any other surface able to provide an appropriate arrangement of apolar groups will strongly bind to the dehydronic region (provided that geometric constraints, i.e. shape complementarity, are satisfied). The completion of the desolvation shell of a structure-determining hydrogen bond has the same significance in understanding protein structure and interactions as completing electron shells has in understanding the periodic table of the elements in chemistry.

Examination of protein-protein interaction interfaces fully bears out the dehydron interpretation. Appropriate complementarity is achieved by over-exposed apolar groups and dehydrons (rather than H-bond acceptors and donors, or positively and negatively ionized residues, although these may play a minor rôle). One also notes that the subunits of haemoglobin, a very stable and soluble (i.e. nonsticky), protein has just three dehydrons: two are at the interface with the other subunits, and one is the bond connecting residues 5 and 8, i.e. flanking the sickle cell anaemia mutation site at residue 6. Furthermore, the prion protein, which is pathologically sticky, has an extraordinarily high density of dehydrons (mean ρ is only about 11).

There are also evolutionary implications. It has long been realized that the evolution of proteins via mutations in their corresponding genes is highly constrained by the need to maintain the web of functional interactions. There is a general tendency for proteins in more evolved species to be able to participate in more interactions; they have more dehydrons. For example, mollusc myoglobin is a perfectly wrapped protein and functions as a loner.

---

[5]Empirically, a certain threshold density of dehydrons per unit area should be exceeded for a surface to qualify as sticky.

Whale myoglobin is in an intermediate position, and human myoglobin is poorly wrapped, hence sticky, and operates together with other proteins as a team. Although folds in a protein of a given function are conserved as species diverge, wrapping is not (even though the sequence homology might still be as much as 30%). Structural integrity becomes progressively more reliant on the interactive context as a species becomes more advanced.

A corollary is that the proteins of more complex species are also more vunerable to fall into pathological states. The prion diseases form a good example: they are unknown in microbes and lower animals. Moreover, they mainly attack the brain, the most sophisticated and complex organ in the living world.

The dehydron concept can also be used to fold a peptide chain *ab initio*.

10.5.3 Experimental techniques for protein structure determination

High throughput methodology (also called structural genomics) comprises the following steps:

1. Select the gene for the protein of interest;

2. Make the corresponding cDNA;

3. Insert the cDNA into an expression system;

4. Grow large volumes of the protein in culture (if necessary with appropriate isotopic labelling of C and N);

5. Purify the protein (using affinity chromatography);

6. Crystallize the protein (usually in unusual salt conditions) and record the X-ray diffractogram,[6] or carry out nuclear magnetic resonance spectroscopy (one or more of ^{1}H, ^{13}C, ^{15}N) to yield the pattern of through-bond and through-space couplings, with a fairly concentrated solution of the protein.

7. Calculate the atomic coordinates;

8. Refine the structure by minimizing interatomic potentials, or use Ramachandran plots.

---

[6]Multiple isomorphous replacement—MIR—whereby a few heavy atoms are introduced into the protein, which is then remeasured, is used to determine the diffraction phases. The heavy atoms should not, of course, induce any changes in the protein structure.

Under favourable conditions, X-ray diffraction and n.m.r. spectroscopy can yield structures at a resolution of 1 Å. Some of the difficulties in these procedures are:

1. The protein may not crystallize. Membrane proteins are especially problematical, but their structures may be obtainable from high-resolution electron diffraction of two dimensional arrays, or by crystallizing them in a cubic phase lipid;

2. Hydrogen atoms are insufficiently electron dense to be registered in the X-ray diffractogram (but are detectable in the experimentally more onerous neutron diffraction);

3. Energy refinement will yield the majority structure. Most proteins have two or more stable structures, which may be present simultaneously, although in unequal proportions;

4. The crystal structure, or the structure in concentrated solution, may not be representative of the native structure(s);

5. N.m.r. cannot cope with large proteins (the spectra become too complicated, and the assignment of peaks to the individual amino acids along the sequence becomes problematical);

6. N.m.r. yields a set of distance constraints, but there are usually so many that the problem is overdetermined, and no physically possible structure can satisfy all of them.

Protein stability can be assessed by determining the structure of a protein at different temperatures. Since thermal denaturation is accompanied by a large change in specific heat, whose midpoint provides a quantitative parameter characterizing stability, microcalorimetry is a useful technique for assessing stability.

10.5.4   Protein structure overview

The techniques described in the previous subsection revealed that proteins have a compact structure akin to a ribbon folded back and forth. Drop a piece of thick string about a metre long on a table, pick is up and push it together between one's hands. This gives a fair impression of typical protein structure. α-helices and β sheets are called secondary structure (the primary structure is the sequence of amino acids). The arrangement of secondary structure elements is called the tertiary structure. Quaternary structure denotes arrangements of individual folded peptide chains to form supramolecular complexes. Quinary structure is the network of other proteins with which a protein interacts.

The number of basic shapes in which proteins fold (i.e. the variety of tertiary structures) seems to be far smaller ($\sim 10^4$) than the number of possible sequences. Individual examples of sequences with less than 10% homology folding into essentially the same structure are known. Moreover some folds are very common, whereas others are rare.

10.6 Polysaccharides

Monosaccharides (sugars) are carbohydrates whose chemical composition is give by the empirical formula $(CH_2O)_n$, with typically $n = 3, 4, 5$, and 6. They are linked together via one of their oxygen atoms in an ether-like linkage to form oligomers and polymers. Saccharide monomers have many -OH groups, and there is much variety in their choice for linking. Some oligosaccharides are metabolic intermediates; they are very often used to modify proteins and lipids, with profound influence on their structure and reactivity.[7] For example, if one sugar is missing from transferrin, an iron-transporting protein in the blood with several glycosylated amino acids, the bearer has an abnormal skin colour, liver problems etc. Oligosaccharides are extensively used to confer specificity of binding e.g. in the immune system. Longer polysaccharides are used to store energy and as structural components, etc. Their assembly is not templated, but accomplished by enzymes. There is considerable variety in the sequence of nominally identical heterooligosaccharides.

Cellulose is a long unbranched chain of glucose monomers linked head to tail. As the major constituent of plant cell walls, there is probably more cellulose on earth than any other organic material. The chains are packed side by side to form microfibrils, which are typically a mixture of two crystalline forms, I_α and I_β, and whose diameter ranges from about 3 nm in most plants to about 20 nm in sea squirts. The chains are held together by hydrogen bonds.

Problem. Examine whether polysaccharides could be used as the primary information carrier in a cell.

---

[7]See Dwek & Butters for a recent overview.

10.7   Lipids

Lipids are not polymers but in water they spontaneously assemble to form large supramolecular structures (planar bilayer membranes and closed bilayer shells, called vesicles). Lipids are amphiphiles, i.e. they consist of a polar moiety (the 'head') attached to an apolar one (the 'tail', typically an alkane chain). The structures formed when lipids are added to water depend on the relative sizes of the polar and apolar moieties. If the tail is thinner than the head, as with many detergents, micelles, compact spherical aggregates with all the head facing outwards, may form. Natural lipids are typically roughly cylindrical, i.e. the head has about the same diameter as the tail and readily form planar or slightly curved membranes (figure 10.7). Obconical shapes (head larger than tail) favour convex structures of small radius.

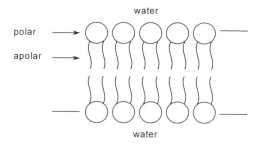

Figure 10.7: A bilayer lipid membrane formed by two apposed sheets of molecules.

A large number of natural lipids are known and found in natural membranes; both the head groups and tails can be varied. A small selection is shown in figure 10.8. The lipid repertoire of a cell or organism is called the 'lipidome'. This diversity allows the shape, fluidity, permeability, affinity for macromolecules etc. of membranes to be adjusted. The biosynthesis of lipids and other membrane components such as cholesterol is of course carried out by enzymes, but the regulation of their abundance and activity is not well understood, and the importance of their variety has probably been underestimated. Most enzymes are attached to membranes and the lipids probably play a far more active rôle than merely functioning as a passive matrix for the protein—which may constitute more than 50% of the membrane. The covalent attachment of a lipid to a protein, typically at a terminal amino acid, is a significant form of post-translational modification.

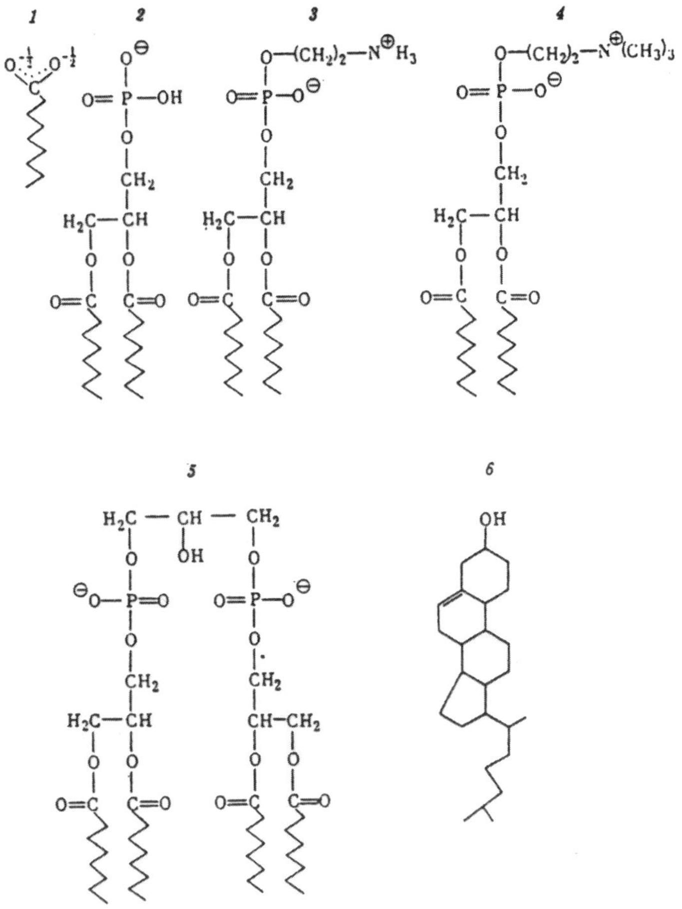

Figure 10.8: Some naturally occurring lipids and membrane components. 1, a fatty acid; 2, phosphatidic acid; 3, phosphatidylethanolamine; 4, phosphatidylcholine; 5, cardiolipin (diphosphatidylglycerol); 6, cholesterol.

Part III

Applications

Chapter 11

Introduction to Part III

Figure 11 is a simplified version of figure 9.1 and highlights the principle objects of investigation of bioinformatics. The field could be said to have begun with individual gene (and hence protein) sequences; typical problems addressed were the extraction of phylogenies from comparing sequences of the same protein over a wide range of different species, and the identification of a gene of unknown function by comparison with the knowledge base of sequences of known function, via the inferential route:

sequence homology ⇒ structural homology ⇒ functional homology. (11.1)

There are, however, plenty of examples of structurally similar proteins with different sequences, or functionally different proteins with similar structures etc. Associated with these endeavours were technical problems of setting up and maintaining databases of sequences and structures.

The bioinformatics landscape was dramatically transformed by the availability of whole genomes, and, at roughly the same time (although there was no especial connexion between the developments), whole proteomes and whole metabolomes. Far wider ranging comparisons could now be carried out; in particular a global vision of regulation seemed to be within grasp. Part III focuses on these developments; table 11.1 recalls the magnitude, at the level of the raw materials, of the problems to be solved.

Genomics is concerned with the analysis of gene sequences, and there are two main territories of this work: (1) comparison of gene sequences, and (2) analysis of the succession of symbols in sequences. The first attempts to elucidate the function of sequences whose function is unknown by comparing the "unknown" sequence with sequences of known function. It is based on the principles (cf. 11.1) that similar sequences encode similar protein structures, and similar structures encode similar functions (there are however many examples for which these principles do not hold). One also compares sequences

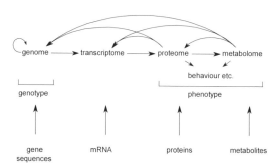

Figure 11.1: The relation between genes, mRNA, proteins and metabolites. The curved arrows in the upper half of the diagram denote regulatory processes.

Object	number
cell types	250
genes	30 000
mRNA	10^5
proteins[a]	3×10^5
expressed proteins[b]	10^3–10^4

Table 11.1: Approximate numbers of different molecular objects in the human body.
[a] Potential repertoire. [b] In a given cell type.

known to code for the same protein (functionally speaking) in different organisms, in order to deduce phylogenetic relationships. A further branch compares the sequences of healthy and diseased organisms, in an attempt to assign genetic causes to disease. The second attempts to assign function to a gene by searching for regularities (the "grammar" of the sequence). In its purest form, genomics could be viewed sinply as the study of the nonrandomness of DNA sequences. This endeavour is still inchoate, since the regularities and their relation to function are not understood. One may, however, be able to predict the structure from the sequence, which can then be used to advance the search for function. The term 'structural genomics' denotes the assignment of structure to a gene product by any means available; 'functional genomics' refers to the assignment of function to a gene product. Proteomics focuses on gene products, i.e. proteins. The primary task is to correlate the pattern of gene expression with the state of the organism. For any given cell, typically only 10% of the genes are actually translated into proteins under a given set of conditions and at a particular epoch in the cell's life. On the other hand, a given gene sequence can give rise to tens of different proteins, by varying the arrangements of the exons and by post-translational modification. Insofar as proteins are the primary vehicle of phenotype, proteomics constitutes a bridge, or communication channel, between genotype and phenotype. One may think of the proteome as the "vocabulary" of the genome: just as we use words to convey ideas and build up our individual characters, so is the genome helpless without proteins. Clearly the proteome forms the molecular core of epigenetics. Once the expression data is available, work can start on its analysis. Via the proteome, genetic regulatory networks can be elucidated.

The primary raw data of proteomics comes from either the transcriptome—a list of all the transcribed mRNAs and their abundances at a particular epoch (the transcriptome)—or the proteome—a list of all the translated proteins and their abundances, or net rates of synthesis, at a particular epoch (the proteome). It is not surprising that there are often huge differences between the transcriptome and proteome. Experimentally, the compiling of such a list involves separating the proteins from one another, and then identifying them.

Comparison between the proteomes of diseased and healthy organisms forms the foundation of the molecular diagnosis of disease.

An important division of proteomics deals with the interactions between proteins. It is indeed so important that a special word has been given to it—interactomics. The primary raw data of interactomics is a list of the affinities of each protein with every other protein in the cell, as well as non-protein material such as lipid bilayers and polysaccharides, and DNA and RNA of course.

Another division is called glycomics—the investigation of protein glycosylation.

Computational proteomics refers to the study of entire proteomes using the genome, looking e.g. for structural features such as transmembrane helices. Here the computational approach is especially important since proteins embedded in lipid membranes by three or more transmembrane helices are very poorly recovered by current methods of experimental proteomics.

The investigation of protein products is called metabolomics. The metabolome comprises all the molecules apart from proteins and DNA (lipids and polysaccharides are also usually excluded) in the cell, and metabolomics is concerned with their identification, abundances, and localization.

Chapter 12

Genomics

We start with a couple of definitions: the genome is the ensemble of genes in an organism, and genomics is the study of the genome. The major goal of genomics is to determine the function of each gene in the genome, i.e. annotate the sequence. This is sometimes called functional genomics. Figure 12.1 gives an outline of the topic. The starting point is the gene; we shall not deal with gene mapping, since it is already well covered in genetics textbooks. We shall view the primary experimental data of genomics as the actual nucleotide sequence, and reiterate that genomics could simply be viewed as the study of the nonrandomness of DNA sequences.

The first section of this chapter will briefly review gene sequencing. The next essential step is to identify where are the genes, and promoter and other sequences possibly involved in regulation—in brief, all biochemically active sites—since even a minimal phenotype must include the regulatory network controlling expression and activity, as well as a list of the expressed genes.

One the coding sequences, i.e. the genes, have been identified, in principle one can determine the protein structure from the sequence alone. This has already been covered in §10.5.2. Once structure is available function might be deduced; there is no general algorithm for doing so, but comparison with proteins of known function whose structure is already known may help to elucidate the function of new genes.

The comparison of sequences of genes coding for the same (functionally speaking) protein in different species is the basis for constructing molecular phylogenies.

In practice, the comparison of sequences of unknown function with sequences of known function has turned out to be as useful for functional genomics as for molecular phylogenies: in other words, it is not apparently necessary to pass by the intermediate step of structure in order to deduce function of a gene, or at least to be able to make a good guess about it.

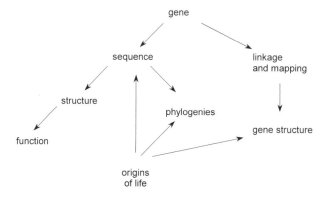

Figure 12.1: The major parts of genomics and their interrelationships. The passage from sequence to function can bypass structure via comparison with sequences of known structure.

12.1 DNA sequencing

12.1.1 Extraction of nucleic acids

The following steps are typical of what is required:

1. Cell separation from the medium in which they are grown by filtration or centrifugation;

2. Cell lysis, i.e. disrupting the cell membranes, mechanically or with detergent, enzymes etc., and elimination of cell debris;

3. Isolation of the nucleic acids by selective adsorption followed by washing and elution,[1]

12.1.2 The polymerase chain reaction

If the amount of DNA is very small, it can be multiply copied ('amplified') by the polymerase chain reaction (PCR) before further analysis. The following steps are involved:

1. Denature (separate) the two strands at 95 °C;

---

[1]This procedure may yield a preparation containing RNA as well as DNA, but RNA binds preferentially to boronate and thus can be separated from DNA.

2. Lower the temperature to 60 °C and add primer, i.e. short synthetic chains of DNA which bind at the beginning (3′ end) of the sequence to be amplified;

3. Add DNA polymerase (usually extracted from thermophilic *Thermus aquaticus* and hence called Taq polymerase) and deoxyribonucleose triphosphates (dNTPs, i.e. an adequate supply of monomers); the polymerase synthesizes the complementary strand starting from the primer;

4. Stop DNA synthesis (e.g. by adding an auxiliary primer complementary to the end of the section of the template to be copied; go to step 1.

The concentration of single strands doubles on each cycle up to about twenty repetitions, after which it declines. Miniature lab-on-a-chip devices are now available for PCR.

12.1.3 Sequencing

The classical technique is that devised by Sanger. One starts with many single-stranded copies of the unknown sequence, to which a known short marker sequence has been joined at one end. An oligonucleotide primer complementary to the marker is added, together with DNA polymerase and nucleotides. A small proportion of the nucleotides are fluorescently labelled dideoxynucleotides lacking the hydroxyl group necessary for chain extension. Hybridization of the primer to the marker initiates DNA polymerization templated by the unknown sequence. Whenever one of the dideoxynucleotides is incorporated, extension of that chain is terminated. After the system has been allowed to run for a time, such that all possible lengths have been synthesized, the DNA is separated into single strands and separated electrophoretically on a gel. The electropherogram shows successive peaks differing in size by one nucleotide. Since the dideoxynucleotides are labelled with a different fluorophore for each base, the successive nucleotides in the unknown sequence can be read off by observing the fluorescence of the consecutive peaks.

A useful approach for very long unknown sequences (such as whole genomes) is to randomly fragment the entire genome (e.g. using ultrasound). The fragments, approximately 2 megabases long and sufficient to cover the genome five to tenfold, are cloned into a plasmid vector,[2] inserted into a bacterial genome and multiplied. The extracted and purified DNA fragments are then sequenced as above. The presence of overlaps allows the original

---

[2]In this context, 'vector' is used in the sense of vehicle.

sequence to be reconstructed.[3] This method is usually called shotgun sequencing. Overlaps are not of course guaranteed, but gaps can be filled in principle by conventional sequencing.[4]

Every aspect of sequencing (reagents, procedures, separation methods etc.) has of course been subject to much development and improvement since its invention (in Sanger's original method, the dideoxynucleotides were radioactively labelled), and there are now high-throughput automated methods in routine use.

Another popular technique is pyrosequencing, whereby one kind of nucleotide only is added to the polymerizing complementary chain; if it is complementary to the unknown sequence at the actual position, pyrophosphate is released upon incorporation of the complementary nucleotide. Using some other reagents, this is converted to ATP, which is then hydrolysed by the chemiluminescent enzyme luciferin. The technique is also suitable for automation.

New techniques are constantly being developed, with especial interest being shown in single molecule sequencing, which would obviate the need for amplication of the unknown DNA.[5] One should also note inexpensive methods designed to detect the presence of a mutation in a sequence. Steady automation has enabled larger and larger pieces of DNA to be tackled.

12.1.4   Expressed sequence tags

Expressed sequence tags are derived from the cDNA complementary to mRNA. They consist of the sequence of typically 200–600 bases of a gene, sufficient to uniquely identify the gene. The importance of ESTs is tending to diminish as sequencing methods get more powerful.

ESTs are generated by isolating the mRNA from a particular cell line or tissue and reverse-transcribing it into cDNA, which is then cloned into a vector to make a "library".[6] Some 400 bases from the ends of individual clones are then sequenced.

If they overlap, ESTs can be used to reconstruct the whole sequence as in shotgun sequencing, but their primary use is to facilitate the rapid identification of DNA. For various reasons, the sequences are typically considerably less reliable than those generated by conventional gene sequencing.

---

[3]This is somewhat related to Kruskal's multidimensional scaling (MD-SCAL) analysis.

[4]Unambiguously assembled nonoverlapping sequences are called 'contigs'.

[5]See França et al. for a review, and Braslavsky et al. for a recent single molecule technique.

[6]In this context, 'library' is used merely to denote 'collection'.

12.2 Gene identification

The ultimate goal of gene identification is automatic annotation: to identify all biochemically active portions of the genome by algorithmically processing the sequence, and predict the reactions and reaction products of those portions coding for proteins. At presence we are still some way from this goal. Success will not only allow one to discover the function of natural genes, but should also enable the biochemistry of new, artificial sequences to be predicted, and ultimately to prescribe the sequence necessary to accomplish a given function.

In eukaryotes the complicated exon-intron structure of the genome makes it particularly difficult to predict the course of the key operations of transcription, splicing and translation from sequence alone (eben without the possibility that essential instructions encoded in acylation of histones etc. are transmitted epigenetically from generation to generation).

At present one would be glad for progress in identifying the exons, introns, promoters etc. in each stretch of DNA such that the exons could be grouped into genes and the promoters assigned to genes or groups of genes. This task is usually called gene prediction. One also wishes to identify the genes (in humans, mammals etc.) believed to originate from viruses, and to localize hypervariable regions (e.g. those coding for immunoglobulins). A more ambitious aim is to be able to understand the relationships between the various elements of the gene.

Gene prediction can be divided into intrinsic (template) and extrinsic (lookup) methods. The former are the best candidates for leading to fundamental insight into how the gene works; if they are successful they should then inevitably provide the means to generalize from the biochemistry of natural sequences to yield rules for designing new genes (and genomes) to fulfil specified functions. We shall begin by considering extrinsic methods.

12.3 Extrinsic methods

The principle of the extrinsic or lookup method is to identify a gene by finding a sufficiently similar known object in existing databases. Hence the method is based on sequence similarity (to be discussed in the next section, §12.4), using the still relatively small core of genes identified by classical genetic and molecular biological studies to prime the comparison, i.e. a gene of unknown function is compared with the database of sequences with known function. This approach reflects a widely used, but not necessarily correct (or genuinely useful), assumption that similar sequences have similar functionality.[7] A major limitation of this approach is the fact that at present about a third of the sequences of newly sequenced organisms turn out to match no sufficiently similar known sequences in existing databanks. Errors in the sequences deposited in databases can be a serious problem.

12.3.1   Database reliability

An inference, especially a deductive one, drawn from data is only as good as the data from which it is formed. The huge collections of gene and protein data now available have encouraged the so-called "hypothesis-free" approach, whereby, it is said, "one can make significant discoveries about a biological phenomenon without insight or intuition" (*sic*).[8] This approach is essentially deductive, and hence, in J.S. Mill's view at least, is not likely to be a productive source of new knowledge. The question of the reliability of the data is certainly a matter for legitimate concern. The most pernicious errors are wrong nucleic acid bases in a sequence. The sources of such errors are legion, and range from the usual experimental uncertainty to mistakes in typing the letters into a file on a keyboard. Of course, these errors can be considered as a source of noise, i.e. equivocation, and handled with the ideas developed earlier, especially in Chapter 3. Undoubtedly there is a certain redundancy in the sequences, but these questions of equivocation and redundancy in database sequences and the consequences for deductive inference do not yet seem to have been given the attention they deserve.

12.4   Sequence comparison and alignment

The pairwise comparison of sequences is very widely used in bioinformatics. If it were only a a question of finding matches to more or less lengthy blocks of symbol sequences (e.g. the longest common subsequence, LCS), the task would be relatively straightforward and the main task would be merely to assess the statistical significance of the result, i.e. compare with the null hypothesis that a match occurred by chance (cf. §5.2.1). In reality, however, the two sequences one is trying to compare differ due to mutations, insertions and deletions (cf. §9.3.1), which renders the problem considerably more

---

[7]Note that 'homology' is defined as 'similarity in structure of an organ or molecule, reflecting a common evolutionary origin'. Sequence similarity is insufficient to establish homology, since genomes contain both orthologous (related via common descent) and paralogous (resulting from duplications within the genome) genes.

[8]The sentence, whose origin I forbear to reveal, goes on with "—hence its tremendous popularity."

complicated; one has to allow for gaps, and one tries to make inferences from local alignments between subsequences. A typical example of a fragment of two nucleotide sequences is:

$$\begin{array}{ccccccccccc}
\text{A} & \text{C} & \text{G} & \text{T} & \text{A} & \text{C} & \text{G} & \text{T} & \text{A} & - & \text{G} & \text{T} \\
| & | & & & | & & | & | & | & & | & | \\
\text{A} & \text{C} & - & - & \text{A} & \text{T} & \text{G} & \text{T} & \text{A} & \text{C} & \text{G} & \text{T}
\end{array}$$

where vertical lines indicate matches (matches between gaps ae disallowed), and blanks indicate gaps or mutations. In the absence of gaps, one could simply compute the Hamming distance between two sequences; the introduction of the possibility of gaps introduces two problems: (i) the number of possible alignments becomes very large; and (ii) where are gaps to be placed in sequence space?

If no gaps are allowed, one assigns and sums scores for all possible pairs of aligned substrings within the two sequences to be matched. If gaps are allowed, there are $\binom{2n}{n}$ possible alignments of two sequences each of length n.[9] Even for moderate values of n there are too many possibilities to be enumerated (problem (i), a computational one). It is solved using dynamic programming algorithms. Problem (ii) is solved by devising a scoring system with which gaps and substitutions can be assigned numerical values. Finally one needs to to assess the statistical significance of the alignment. This is still an unsolved problem—let us call it problem (iii).

The essence of sequence alignment is to assign a score, or cost, for each possible alignment: the one with the lowest cost, or highest score, is the best one, and if aligning multiple sequences, degrees of kinship can be assigned on the basis of the score, which has the form

$$\text{total score} = \text{score for aligned pairs} + \text{score for gaps} . \tag{12.1}$$

The score is, in effect, the relative likelihood that a pair of sequences are related. It represents distance, together with the operations (mutations and introduction of gaps) required to edit one sequence onto the other. Sequence alignment attempts to maximize the number of matches while minimizing the number of mutations and gaps required in the editing process. Unfortunately the relative weights of the terms on the right hand side of (12.1) are arbitrary. The main approach to assigning weights to the terms more objectively is to study many extant sequences from organisms one knows from independent evidence to be related. In principle, under a given set of conditions, e.g. a

---

[9]This is obtained by considering the number of ways of intercalating two sequences while preserving the order of symbols in each.

certain level of exposure to cosmic rays, a given mutation presumably has a definite probability of occurrence, i.e. it can, at least in principle, be derived from an objective set of data according to the frequentist interpretation, but the practical difficulties, and the possibility that such probabilities may be specific to the sequence neighbouring the mutation, make this an unpromising approach.

Whereas with DNA sequences, a nucleotide is either matched or is not, with polypeptides a substitution might be sufficiently close chemically so as to be functionally neutral. Hence, if alignments are carried out at the level of amino acids, exact matches and substitutions are dealt with by compiling an empirical table, based on chemical or biological knowledge or both, of degrees of equivalence.[10] There is no uniquely optimal table. To construct one, a good starting point is the table of amino acids (10.6). Isoleucine should have about the same score for substitution by leucine as for an exact match, etc.; substitution of a polar for an apolar group, or lysine for glutamic acid (say) would be given low or negative scores. The biological approach is to look at the frequencies of the different substitutions in pairs of proteins that can be considered to be functionally equivalent from independent evidence (e.g. two enzymes that catalyse the same reaction).

In essence, the entries in a scoring matrix are numbers related to the probability of a residue occurring in an alignment. Typically they are calculated as (the logarithm of) the probability of the "meaningful" occurrence of a pair of residues divided by the probability of random occurrence. Probabilities of "meaningful" occurrences are derived from actual alignments "known to be valid". The inherent circularity of this procedure gives it a temporary and provisional air.

In the case of gaps, the (negative) score might be a single value per gap, or could have two parameters, one for starting a gap, and another, multiplied by the gap length, for continuing it (called an affine gap cost). This takes some slight account of possible correlations in the history of changes presumed to have been responsible for causing the divergence in sequences. The scoring of substitutions considers each mutation to be an independent event, however.

Dynamic programming algorithms

The concept of dynamic programming comes from operations research, where it is commonly used to solve problems that can be divided into stages with a decision required at each stage. A good generic example is the problem of finding the shortest path on a graph. The decisions are where to go next at

---

[10]For example, BLOSUM50, a 20 × 20 score matrix (histidine scores 10 if replacing histidine, glutamine zero, and alanine –3, and so on). The diagonal terms are not equal.

each node. It is characteristic that the decision at one stage transforms that state into a state in the next stage. Once that is done, from the viewpoint of the current state the optimal decision for the remaining states does not depend on the previous states or decisions. Hence it is not necessary to know a node was reached, only that it was reached. A recursive relationship identifies the optimal decision for stage M, given that stage $M+1$ has already been solved; the final stage must be solvable by itself.

The following is a generic dynamic programming algorithm (DPA) for comparing two strings, $S1$ and $S2$ with $M[i,j]$ = cost or score of $S1[1..i]$ and $S2[1..j]$:[11]

```
M[0, 0] = z
for each i in 1 .. S1.length
   M[i,0] = f( M[i-1, 0 ], c(S1[i], "_" ) )       -- Boundary

for each j in 1 .. S2.length
   M[0,j] = f( M[0,  j-1], c("_", S2[j] ) )       -- conditions

for each i in 1 .. S1.length and j in 1 .. S2.length
   M[i,j] = g(f(M[i-1, j-1], c(S1[i], S2[j])),  -- (mis)match
              f(M[i-1, j ], c(S1[i], "_" )),   -- delete S1[i]
              f(M[i,  j-1], c("_",  S2[j])))   -- insert S2[j]
```

Applied to sequence alignment, two varieties of DPA are in use: the Needleman-Wunsch ('global alignment') algorithm that builds up an alignment starting with alignments of small subsequences, and the Smith-Waterman ('local alignment') algorithm that is similar in concept, except that it does not systematically move through the sequences from one end to the other, but compares subsequences anywhere.

It is often tacitly assumed that the sequences are random, i.e. incompressible, but if they are not (i.e. they are compressible to some degree) this should be taken into account.

There are also some heuristic algorithms (e.g. BLAST and FASTA) that are faster than the DPAs. They look for matches of short subsequences, which may be only a few nucleotides or amino acids long, that they then seek to extend. As with the DPAs some kind of scoring system has to be used to quantify matches.

Although sequence alignment has become very popular, some of the assumptions are quite weak and there is strong motivation to seek alternative methods for evaluating the degree of kinship between sequences, not based on

---

[11]Allison et al.

symbol-by-symbol comparison. For example, one could evaluate the mutual information between strings a and b:

$$I(s_a, s_b) = I(s_b, s_a) = I(s_a) - I(s_a|s_b) = I(s_b) - I(s_b|s_a) \ . \qquad (12.2)$$

Multiple alignment is an obvious extension of pairwise alignment.

12.5 Intrinsic methods

The template or intrinsic approach involves constructing concise descriptions of prototype objects, and then identifying genes by searching for matches to such prototypes. An elementary example is searching for motifs (i.e. short subsequences) known to interact with particular drugs. The *motif* is often defined more formally along the lines of a sequence of amino acids that defines a substructure in a protein that can be connected in some way to protein function or structural stability, and hence which appear as conserved regions in a group of evolutionarily related gene sequences. This is not a strong definition, not least because genes are often considered to be evolutionarily related because they share a common sequence. Moreover, the motif concept is really based on a mosaic view of the genome that is opposed to the systems view.

The construction of the concise descriptions could be either deductive or inductive. A difficulty is that extant natural genomes are not elegantly designed from scratch, but assembled *ad hoc*, and refined by 'life experience'. It is hoped that the use of fuzzy criteria may help to overcome this problem.

In practice, this method often boils down to either computing one or more parameters from the sequence and comparing them with the same parameters computed for sequences of known function, or searching for short sequences that experience has shown are characteristic of certain functions.

12.5.1 Statistical regularities

In essence this is a search for certain parameters (sometimes called coding measures) that experience has shown are correlated with coded protein function. Examples are codon usage (the frequencies of each of the 64 codons), amino acid usage, hexamer usage (the frequencies of hexamers), autocorrelation functions of bases, dimers, trimers (i.e. codons) etc., and so on. A major difficulty is that a gene, or an intron, will typically be too short to allow a parameter to be estimated sufficiently precisely to allow an identification to be made.

12.5.2 Signals

In the context of intrinsic methods for assigning function to DNA, the term denotes short sequences relevant to the interaction of the gene expression machinery with the DNA. In effect, one is paralleling the action of the cell (e.g. the transciption, splicing and translation operations) by trying to recognize where the gene expression machinery interacts with DNA. In a sense, therefore, this topic belongs equally well to interactomics. Much use has been made of so-called consensus sequences: these are formed from sequences well-conserved over many species by taking the most common base at each position. The distance (i.e. the Hamming distance) of an unknown sequence is then computed for the unknown sequence; the closer they are, the more likely it is that the unknown sequence has the same function as that represented by the consensus sequence. Useful signals include start and stop codons. More sophisticated signals include sequences predicted to result in unusual DNA bendability, or known to be involved in positioning DNA around histones; intron splice sites in eukaryotic pre-mRNA and sequences corresponding to ribosome binding sites on RNA; etc.

Especial effort has been devoted to identifying promoters, which are of great interest as potential targets for new drugs. It is a hard problem because of large and variable distance between the promoter(s) and the sequence to be transcribed. The approach relies on relatively well conserved sequences (i.e. effectively consensus sequences) such as TATA or CCAAT. Other sites for protein-DNA interactions can be examined in the same way, indeed the entire transcription factor binding site can be included in the prototype object, which allows more sophistication (e.g. some constraints between the sequences of the different parts) to be applied.

12.5.3 Hidden Markov models (HMM)

Knowledge of the actual biological sequence of processing operations can be used to exploit the effect of the constraints on (nucleic acid) sequence that these successive processes imply. One presumes that the Markov binary symbol transition matrices are slightly different for introns, exons, promoters, enhancers, the complementary strand, etc. One constructs a more elaborate automaton, or automaton of automata, in which the outer one controls the transitions between the different types of DNA (introns, exons, etc.) and the inner set gives, for each type, the sixteen different binary transition probabilities for the symbol sequence. More sophisticated models use higher order chains for the symbol

transitions; further levels of automata can also be introduced. The epithet 'hidden' is intended to signify that only transitions from symbol to symbol are observable, not transitions from type to type. The main problem is the statistical inadequacy of the predictions. A promoter may only have two dozen bases; a fourth order Markov chain for nucleotides has of the order of 10^{10} transition probabilities.

12.6   Beyond sequence

Proteomics data (see Chapters 13 and 14) is integrated with sequence information in the attempt to assign function. Proteins whose mRNA levels are correlated, proteins whose homologues are fused into a single gene in some organism, those which have evolved in a correlated fashion, those whose homologues operate together in a metabolic path, or which are known to physically interact, can all be considered to be linked in some way. For example, a protein of unknown function whose expression profile matches that of a protein of known function in another organism is assigned the same function. In a literary analogy, one could rank the frequencies of words in an unknown and known language and assign the same meanings to the same ranks. Whether the syntax of gene expression is sufficiently shared by all organisms to allow this to be done reliably is an open question at present.

Other kinds of data assisting protein function prediction are structure prediction (cf. 11.1), intracellular localization, signal peptide cleavage sites of secreted proteins, glycosylation sites, lipidation sites, phosphorylation sites, other sites for posttranslational modification, cofactor binding sites, dehydron density, etc.

12.7   Phylogenies

The notion that life forms evolved from a single common ancestor, i.e. that the history of life is a tree, is pervasive in biology.[12]

Before gene and protein sequences became available, trees were constructed from the externally observable characteristics of organisms. Each organism is therefore represented by a point in phenotype space. In the simplest (binary) realization, a characteristic is either absent (0) or present (1), or is present in either a primitive (0) or an evolved (1) form. The distance between species, compared in pairs, can by computed as a Hamming distance, i.e. the number of different characteristics. For example, consider

---

[12]The concept of phylogeny was introduced by E. Haeckel.

three species A, B, C to which ten characteristics labelled a to j are assigned:

	a	b	c	d	e	f	g	h	i	j
A	1	1	1	1	1	1	1	1	0	0	1
B	0	0	0	0	0	1	1	1	0	0
C	0	0	0	0	0	0	0	0	1	0

$$(12.3)$$

This yields the distance matrix:

	A	B	C
A	0.0		
B	0.7	0.0	
C	0.9	0.4	0.0

$$(12.4)$$

The species are then clustered: the first cluster is formed from the closest pair, viz. B and C, the next cluster is formed between this pair and the species closest to its two members (and so on in a larger group) to yield the following tree or dendrogram:

$$(12.5)$$

This is the classical method; the root of the tree is the common ancestor.

An alternative method, called cladistics, counts the number of transformations necessary to go from from a primitive to an evolved form. Hence in the example C differs by just one transformation from the putative primitive form (all zeros). Two transformations (of characters f and j) create a common ancestor to A and B, but it must be on a different branch from that of C, which does not have evolved forms of those two characteristics. This approach yields a different tree:

$$(12.6)$$

The principle of construction of molecular phylogenies is to use the sequences of the same genes (i.e. encoding a protein of the same function) in different organisms as the characteristics of the species, i.e. it is based on genotype rather than phenotype. In actual practice, protein sequences are used, which are intermediate between genotype and phenotype. In the earliest studies (1965 ∼ 1975) cytochrome c was a popular object, since it is found in nearly all organisms, from bacteria to man. Later the sequence of the small subunit of ribosomal RNA (rRNA), another essential and universal object, was used.[13] Nowadays one can, in principle, analyse whole genomes.

A chronology can be established on the premiss that the more changes there are, the longer the elapsed time since the species diverged (assuming that the changes occur at a constant rate). This can be criticized since although the unit of change is the nucleotide, selection (the engine of speciation) acts on the amino acid; some nucleotide mutations lead to no change in amino acid due to the degeneracy of the code. There is actually little real evidence that mutations occur at random (i.e. both the site and the type of mutation).

A difficulty with molecular phylogenies is the fact that lateral gene transfer (LGT), especially between bacteria, and between archaea, may vitiate the calculated distances. A plausible counterargument is that rRNA should be unaffected by LGT due to its fundamental place in cell metabolism.

A further difficulty is a computational one: that of finding the optimal tree, since usually one is interested in comparing dozens (and ultimately millions) of species. The basic principle is that of parsimony: one seeks to construct the tree with the least possible number of evolutionary steps. Unfortunately this is an NP-complete problem, hence the computation time grows exponentially with the number of species; even a mere 20 species demands the analysis of almost 10^{22} possible trees!

---

[13]rRNA has been championed by C. Woese.

Chapter 13

Proteomics

The proteome is the ensemble of expressed proteins in a cell, and proteomics is the study of that ensemble, i.e. the identification and determination of amounts, locations and interactions of all the proteins. The tasks of proteomics are summarized in figure 13.1.

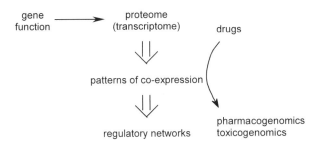

Figure 13.1: The major parts of proteomics and their interrelationships.

We have seen in Chapter 9 how the gene is first transcribed into messenger RNA (mRNA), and a given gene, especially in a eucaryotic cell in which the gene is a mosaic of introns (Is) and exons (E)s, can be assembled to form different mRNAs (e.g. if the gene is $E_1 I E_2 I E_3 I E_4 I E_5$ one could form mRNAs $E_1 E_2 E_3 E_4 E_5$, $E_1 E_3 E_4 E_5$, $E_1 E_3 E_5$ etc.). The ensemble of these transcripts is called the transcriptome, and its study is called transcriptomics. Due to the variety of assembly possibilities, the transcriptome is considerably larger (i.e. contains more types of objects) than the genome.

After the mRNA is translated into a protein, the polypeptide may be modified by:

1. Cutting off a block of amino acids from either end;

189

2. Covalently adding a large chemical group to an amino acid, e.g. a fatty acid or an oligosaccharide;

3. Covalently modifying an amino acid, e.g. by serine or threonine phosphorylation, or acetylation;

4. Oxidizing or reducing an amino acid, e.g. arginine deimination or glutamine deamidation.

Modifications 2 and 3 may well be reversible, i.e. there may be a pool of both modified and unmodified forms in the cell at any instant. More than 200 post-translational modifications (PTM) have been identified. They can significantly change conformation, hence catalytic activity of an enzyme, and inter-molecular specificity, hence binding, localization etc., all crucial aspects of the dynamic system which precedes the phenotype in the genotype → phenotype succession.

These modifications increase the potential repertoire of proteins expressible from each gene typically by 1–2 orders of magnitude (since many combinations are possible) compared with the repertoire of genes. Notice that effecting these modifications requires enzymes, hence the proteome is highly self-referential.

Although the number of different proteins therefore far exceeds the number of genes, the actual number of proteins present in a cell at any one instant may well be much smaller than the number of genes, since only a part of the possible repertoire is likely to be expressed. Each cell type in an organism has a markedly different proteome. The proteome for a given cell type is likely to depend on the environment; unlike the genome, therefore, which is relatively static, the proteome is highly dynamic.

In this chapter we shall cover experimental aspects (sample preparation, separation and identification) and the analysis of the results obtained, for both transcriptomics and proteomics.

Proteomics is often defined so as to encompass what is otherwise called interactomics, the study of the ensemble of molecular interactions, especially protein-protein interactions, in a cell, including those that lead to the formation of more or less long-lived multiprotein complexes. These aspects are covered in the next chapter.

13.1   Transcriptomics

The goal of transcriptomics is to identify and quantify the amounts of all the mRNA in a cell. This is mainly done using microarrays ('gene chips').

The principle of a microarray is to coat a flat surface with spots of DNA complementary to the expressed mRNA, which is then captured because of the complementary base pairing (hybridization) between DNA and RNA (A–U, C–G, G–C, T–A) and identified. The relationship of a microarray to a classical affinity assay resembles that of a massively parallel processor to classical linear processor architecture, in which instructions are executed sequentially. The parent classical assay is the Northern blot.[1]

Microarrays consist of a two dimensional array, typically a few square millimeters in overall area, of more or less contiguous patches, the area of each patch being a few square micrometres (or less), and each patch on the array having a different chemical composition. Typical microarrays are assembled from one type of substance (e.g. nucleic acid oligomers).

In use, the array is flooded with the sample whose composition one is trying to identify. After some time has elapsed the array is scanned to determine which patches have captured something from the sample.[2] It is, of course, essential that each patch should be addressible, in the sense that the composition of each individual patch is known or traceable. Hence a photomicrograph of the array after exposure to the analyte should allow one to determine which substances have been captured from the sample.

Table 13.1 summarizes some features of microarrays.

Application	captors	sample
genomics	ESTs	DNA
transcriptomics	cDNA	mRNA
proteomics	antibodies	proteins
metabolomics	various	various

Table 13.1: Typical features of microarrays.

In detail, the protocol for a microarray assay would typically involve the following steps:

---

[1]Northern blotting allows detection of specific RNA sequences. RNA is fractionated by agarose gel electorphoresis, followed by transfer (blotting) to a membrane support, followed by hybridization with known DNA or RNA probes.

[2]The same principle (cf. Southern blotting compared with Northern) If one is trying to determine whether certain genes are present in a bacterial culture (for example), the array should be coated with patches of complementary nucleic acid sequences. The DNA is extracted from the bacteria, subjected to some rudimentary purification, separated into single strands, and usually cut into fragments using restriction enzymes, and then poured over the microarray.

Array preparation The chip must be designed on the basis of what one is looking for. Each gene of interest should be represented by at least one, or preferably more, unique subsequences. Once the set of sequences has been selected, there are two main approaches to transfer them to the chip:

1. Assemble successive monomers using microfabrication technology. For example[3] photoactivatable nucleic acid monomers are prepared. Masks, or a laser scanner, activate those patches selected to receive, say, G. After exposure to light, the array is then flooded with G. Then the array is exposed to a different pattern, and again flooded (with a different base), and so on. This technology is practicable up to about 20 cycles, and is highly appropriate wherever linear heterooligomers sharing a common chemistry are required.

2. For all other cases, minute amounts of the receptor substances can be directly deposited on the array, e.g. using laboratory microrobotics technology. This is suitable for large macromolecules, such as proteins, or sets of molecules substances not sharing a common chemstry, or longer oligopeptides. Solutions of the different substances are applied using a type of inkjet technology.

In both cases, each patch can be uniquely identified by its Cartesian array coordinate.

Sample preparation The raw material is processed to release or uncomplex the analytes of interest, and possibly partially purified. The mRNA is typically used to generate a set of complementary DNA molecules (cDNA), which may be tagged with a fluorescent or other kind of label.

Array exposure The array is flooded with the sample and allowed to reach equilibrium.

Then, all unbound sample is washed away. The tagging can be carried out on the chip after removing the unbound molecules, which has the advantage of eliminating the possibility of the tag interfering with the binding.

Array reading The array is scanned to determine which patches have captured molecules from the sample. If the sample molecules have been labelled with a fluorophore, then fluorescent patches indicate binding,

---

[3]Fodor et al.

with the intensity of fluorescence giving some indication of the amount of material bound, which in turn should be proportional to the amount of mRNA expressed in the original sample. The binding of unlabelled samples may be detected by certain compounds which fluoresce when intercalated into double stranded DNA, i.e. the spots hybridized by capture from the sample.

Image processing The main task is to normalize the fluorescent (or other) intensities. It is important when comparing the transcriptomes from two samples (e.g. taken from the same tissue subject to two different growth conditions). A straightforward procedure is to assume that the total amount of expressed mRNA is the same in both cases (which may not be warranted, of course) and to divide the intensity of each individual spot by the sum of all intensities. If the transcriptomes have been labelled with different fluorophores and exposed simultaneously to the same chip, then normalization corrects for differences in fluorescence quantum yield, etc.

Analysis The procedures followed for supervised hypothesis testing will depend on the details of the hypothesis. Very commonly, however, unsupervised exploratory analysis of the results is carried out. This used no prior knowledge, but explores the data on the basis of correlations and similarities. One goal is to find groups of genes that have correlated expression profiles, from which might be inferred that they participate in the same biological process. Another goal is to group tissues according to their gene expression profiles: it might be inferred that tissues with the same or similar expression profile belong to the same clinical state.

Pattern recognition techniquesare used to analyse the data. If a set of experiments comprising samples prepared from cells grown under m different conditions has been carried out, then the set of normalized intensities (i.e. transcript abundances) for each experiment defines a point in m-dimensional expression space, whose coordinates give the (normalized) expressions. Distances between the points can be calculated by e.g. the Euclidean distance metric, i.e.

$$d = \left[\sum_{i=1}^{m} (a_i - b_i)^2 \right]^{1/2} \tag{13.1}$$

for two samples a and b subjected to m different conditions. Clustering algorithms can then be used to group transcripts. The hierarchical clustering procedure is the same as that used to constuct phylogenies (§12.7), i.e. the closest pair of transcripts forms the first cluster, the transcript with the closest mean distance to the first cluster forms the second cluster, and so on. This is the unweighted pair-group method average (UPGMA); variants include single linkage clustering, in which the distance between two clusters is calculated as the minimum distance between any members of the two clusters, etc.

Fuzzy clustering algorithms may be more successful than the above "hard" schemes for large and complex data sets. Fuzzy schemes allow points to belong to more than one cluster. The degree of membership is defined by

$$u_{r,s} = 1/\sum_{j=1}^{m} \left(\frac{d(x_r,\theta_s)}{d(x_r,\theta_j)}\right)^{1/(q-1)} \qquad r = 1,\ldots,N; \quad s = 1,\ldots,m$$

(13.2)

for N points and m clusters (m is given at the start of the algorithm), where $d(x_i,\theta_j)$ is the distance between the point x_i and the cluster represented by θ_j, and $q > 1$ is the fuzzifying parameter. The cost function

$$\sum_{i=1}^{N}\sum_{j=1}^{m} u_{r,s}^{j} d(x_i,\theta_j) ,$$

(13.3)

is minimized (subject to the condition that the $u_{i,j}$ sum to unity) and clustering converges to cluster centres corresponding to local minima or saddle points of the cost function. The procedure is typically repeated for increasing numbers of clusters until some criterion for clustering quality, e.g. the partition coefficient

$$(1/N)\sum_{i=1}^{N}\sum_{j=1}^{m} u_{i,j}^{2} ,$$

(13.4)

becomes stable. The closer the partition coefficient is to unity, the "harder" (i.e. better separated) the clustering.

Instead of using a clustering approach, the dimensionality of expression space can be reduced by principal component analysis (PCA), in which the original dataset is projected onto a small number of orthogonal axes. The original axes are rotated until there is maximum variation of the points along one direction. This becomes the first principal component. The second is the axis along which there is maximal residual variation, and so on.

Microarrays have some limitations, and one should note the following potential sources of problems: manufacturing reproducibility; variation in how the experiments are carried out (exposure duration, temperature gradients, flow conditions, etc., all of which may severely affect the actual amounts hybridized); ambiguity between pre- and post-processed (spliced) mRNA; mRNA fragment size distribution not matching that of the probes; quantitative interpretation of the data; expense. Other techniques have being developed, such as serial analysis of gene expression (SAGE). In this technique, a short but unique sequence tag is generated from the mRNA of each gene using the PCR and joined together ('concatemerized'). The concatemer is then sequenced. The representation of each tag in the sequence will be proportional to the degree of gene expression.

Problem.   How many n-mers are needed to unambiguously identify g genes?

13.2   Proteomic analysis

The proteome can be accessed directly by measuring the expression levels, not of the transcripts, but of the proteins into which they are translated. Not surprisingly, in the relatively few cases for which comparative data for both the transcriptome and proteome have been obtained, the amounts of the RNAs and corresponding proteins are very different, even if all the different proteins derived from the same RNA are grouped together. Translation is an important arena for regulating protein synthesis. Before this became apparent, transcriptomics acquired importance because technically it is much easier to obtain the transcriptome using a microarray than it is to obtain the proteome using laborious two dimensional gel electrophoresis, for example. It was hoped that the transcriptome would be a reasonably faithful mirror of the proteome. This is definitely not the case however: there is no presently discernible unique relationship between the abundance of mRNA and abundance of the corresponding protein. Hence the transcriptome has lost some of its importance; it is "merely" an intermediate stage and does not contribute directly to phenotype in the way that the proteome does. Furthermore, the transcriptome contains no information about the very numerous posttranslational modifications of proteins. On the other hand, to understand the relation between transcriptome and proteome would be a considerable advance in understanding the overall mechanism of the living cell. At present, given that both transcriptome and proteome spaces each have such a high dimension, deducing a relation between trajectories in each is a rather forlorn hope.

The first step in proteomics is to separate all the expressed proteins from each other such that they can be individually quantified, i.e. characterized by type and number. Prior to that, however, the ensemble of proteins have to be separated from the rest of the cellular components. Cells are lysed, proteins are solubilized and cellular debris is centrifuged down. Nucleic acids and lipids are removed, and sometimes very abundant proteins (such as albumin from serum). A subset of proteins may be labelled at this stage, to assist later identification.

A particularly useful form of labelling is to briefly (for 30–40 minutes) feed the living cells with radioactive amino acids (^{35}S-cysteine and methionine are suitable), followed by an abundance of non-radioactive amino acids. The degree of incorporation of radioactivity into the proteins is then proportional to the net rate of synthesis (i.e. biosynthesis rate minus degradation rate).

The two main techniques for separating the proteins in this complex mixture (which is likely to contain several hundred to several thousand different proteins) are:

1. Two dimensional gel electrophoresis (2DGE);

2. Enzymatic proteolysis into shorter peptides followed by column chromatography. Trypsin is usually used as the proteolytic enzyme (protease) since it cuts at well-defined positions (lysines).

The protein mixture may be pretreated (prefractionated), using chromatography or electrophoresis, before proceeding to the separation step, in order to selectively enrich it with certain types of proteins.

Problem. List and discuss the differences between mRNA and protein abundances.

13.2.1 Two dimensional gel electrophoresis (2DGE)

In order to understand the principles of protein separation by 2DGE, let us first recall some of the physico-chemical attributes of proteins. Two important ones are

1. Molecular weight M_r;

2. Net electrostatic charge Z (as a function of pH—the pH at which $Z = 0$ is important as a characteristic parameter).[4]

---

[4]This is known as isoelectric point (i.e.p.), or pI, or point of zero charge (p.z.c.).

Both can be calculated from amino acid sequence (assuming no postransla-tional modifications), from M_r and Z of the individual amino acids. M_r is easy; to calculate Z one has to make the quite reliable assumption that all the ionizable residues are on the protein surface. The calculation is not quite as simple as adding up all the surface charges, since they mutually affect each other (cf. the surface of a silicate mineral: not every hydroxyl group is ionized, even at extremely low pH).[5]

The technique itself was independently developed by Klose and O'Farrell in 1975. The concept depends on the fact that separation by isoelectric point (i.e.p.) is insufficient to separate such a large number of proteins, many of whose i.e.p.s are clustered together. Equally, there are many proteins with similar molecular masses. By applying the two techniques sequentially, however, they can be separated, especially if large (30×40 cm) gels are used.

Proteins in the crude cell extract are dispersed in an aqueous medium con-taining the anionic detergent sodium dodecyl sulphate (SDS) which breaks all noncovalent bonds, i.e. subunits are dissociated, and probably denatured too; the first separation takes place according to isoelectric point by elec-trophoresis on a gel along which a pH gradient has been established; the partly separated proteins are then transferred to a second, polyacrylamide, gel within which separation is effected according to size, i.e. molecular weight if all proteins are assumed to have the same density.

If the cells have been pulse radiolabelled prior to making the extract, then the final gel can be scanned autoradiographically and the density of each spot is proportional to the net rate of protein synthesis. Alternatively (or in par-allel) the proteins can be stained and the gel scanned with a densitometer; the spot density is then proportional to protein abundance. There are some caveats: membrane proteins with more than two transmembrane sequences are poorly recovered by the technique; if ^{35}S met/cys is used, one should note that not all proteins contain the same number of met and cys (but this number is only very weakly correlated with molecular weight); autoradio-graphy may underestimate the density of weak spots, due to low intensity reciprocity failure of the photographic (silver halide) film used to record the presence of the radionucleides; the commonly used Coomassie blue does not stain all proteins evenly; although the unevenness appears to be random and hence should not impose any systematic distortion on the data; rare pro-teins may not be detected at all; several abundant proteins clustered close together may not be distinguishable from each other; and very small and very

---

[5]Linderstrøm-Lang worked out a method of taking these correlations into account; his formula works practically as well as more sophisticated approaches (including explicit numerical simulation by Brownian dynamics) and is much simpler and more convenient to calculate (see Ramsden et al. (1995) for an application example).

large proteins, and those with isoelectric points (pI) at the extremes of the
pH range, will not be properly separated. The molecular mass and isoelectric point ranges are limited by practical considerations. Typical ranges are
$15000 < M_r < 90000$ and $3 < \text{pI} < 8$. Hence the mostly basic (pI typically in
the range 10–14) 50–70 ribosomal proteins will not be captured, as a notable
example (on the other hand these proteins are not supposed to vary much
from cell to cell, regardless of conditions, since they are essential proteins for
all cells, hence they are not considered to be especially characteristic of a
particular cell or metabolic state). Figure 13.2 shows a typical result.

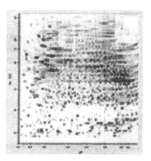

Figure 13.2: A two dimensional gel after staining.

13.2.2   Column chromatography

The principle of this method is to functionalize a stationary solid phase
(granules of silica, for example) packed in a column and pass the sample
(suspended or dissolved in the liquid mobile phase) through it. The func-
tionalization is such that the proteins of interest are bound to the granules,
and everything else passes through. A change in the liquid phase composition then releases the bound proteins. Better separations can be achieved
by "multidimensional" liquid chromatography (MDLC), in which a cation
exchange column (for example) is followed by a reverse phase column. The
number of "dimensions" can be increased further. Usually the technique is
used to prepurify a sample, but in principle, using differential elution (i.e.
many proteins of interest are bound, and then released sequentially by slowly
increasing pH, or the polarity of the liquid), high resolution separations may
also be accomplished. Miniaturization (nano-liquid chromatography) offers
promise in this regard. The output from the chromatography may be fed
directly into a mass spectroscope.

13.2.3   Other kinds of electrophoresis

Free fluid electrophoresis (FFE) is distinguished from chromatography in that there is no stationary phase, i.e. no transport of analytes through a solid matrix such as a gel. The separation medium and the analytes are carried between electrodes, arranged such that the electric field is orthogonal to the flow of the separation medium.[6]

13.3   Protein identification

Two dimensional gel electrophoresis is very convenient since it creates a physical map of the cell's proteins in M_r–i.e.p. space, from which the proteins at given coordinates can actually be cut out and analysed. Hence it is possible to apply Edman sequencing,[7] at least to the more abundant proteins, or Stark C-terminal degradation. The most widely applied technique is based on mass spectrometry (MS), however. It is capable of much higher throughput, and post-translational modifications can be readily detected. Mass spectrometers consist of an ion source, a mass analyser (ion trap, quadripole, time of flight or ion cyclotron) and a detector

The objects to be analysed have to be introduced into the mass spectrometer in the gas phase. This can be achieved by electrospraying or laser desorption ionization. In electrospraying, the proteins are dissolved in salt-free water, typically containing some organic solvent and forced to emerge as droplets from the end of an electrostatically charged silica capillary. As the solvent evaporates the electrostatic charge density increases until the droplets explode. The solution dilution should be such that each protein is then isolated from its congeners. The remaining solvent evaporates and the protein molecules pass into the mass spectrometer. At this stage each protein molecule is typically multiply charged. Sequential quadrupole filters, typically three, are used to achieve adequate discrimination. The mass spectrum for an individual protein consists of a series of peaks corresponding to m/z ratios whose charge z differs by one electron. The middle quadrupole may contain a collision gas (e.g. A) to fragment the protein into smaller peptides.

In laser desorption ionization, usually called MALDI (matrix-assisted laser desorption ionization) or SELDI (surface-enhanced laser desorption/ion-

---

[6]See Patel & Weber for a review.

[7]The N-terminal of the protein is derivatized with phenylisothiocyanate to form a phenylthiocarbamate peptide, and the first amino acid is cleaved by strong acid resulting in its anilothiazolinone derivative plus the protein minus its first N-terminal amino acid. The anilothiazolinone derivative is converted to the stabler phenylthiohydantoin for subsequent HPLC identification.

ization), the protein is mixed with an aromatic organic molecule, e.g. sinapinic acid $((CH_3O)_2OHC_6H_2(CH_2)_2COOH)$, spread out as a thin film, and
irradiated by a pulsed u.v. laser. The sinapinic acid absorbs the light and
evaporates, taking the proteins with it. Other matrices can be used with
infrared lasers.[8] The proteins are typically singly charged, and a time of
flight (ToF) mass spectrometer detects all the ions according to their mass.
MALDI-ToF cannot detect as wide a range of proteins as quadrupole MS,
and the matrix can exert unpredictable effects on the results. Nevertheless,
the vision of spots on a 2D gel being rapidly and sequentially vaporized by
a scanning laser and immediately analysed in the mass spectrometer offers
hope for the development of high throughput proteomics analysis tools.

Newer developments in the field include the application of sophisticated
ion cyclotron resonance mass spectrometers, the use of Fourier transform
techniques, and miniature instrumentation according to the lab-on-a-chip
concept.

MS is also used to characterize the peptide fragments resulting from proteolysis followed by chromatography. Proteins separated by 2DGE can also
be cleaved using trypsin or another protease to yield fragments which as then
mass-fingerprinted using MS.

The proteolytic peptide fragments are encoded as a set of numbers corresponding to their masses, and these numbers are compared with a database
assembled from the mass-fingerprints from known peptides.

13.4   Isotope coded affinity tags (ICAT)

The technique[9] is particularly useful for comparing the expression levels of
proteins in samples from two different sources, for example cells before and
after treatment with a chemical. It is a way of reducing the variety (number of
proteins that have to be separated) of a complex mixture. Proteins from the
two sources are reacted with light and heavy ICAT reagents in the presence
of a reducing agent. The reagents comprise a biotin moiety, a sulfhydryl-
specific iodoacetate moiety and a linker that carries eight ^1H (light) or ^2H
(heavy) atoms. They specifically tag cysteinyl residues on the proteins. The
two batches are then mixed and the proteins cleaved using trypsin. The
fragments, only about a fifth of which contain cysteine, can be readily separated by chromatography on an avidin affinity column (which binds to the
biotin), and finally analysed by mass spectrometry. Singly charged peptides

---

[8]See Chem. Rev. 103 (2003), no 2.

[9]Developed by Aebersold (Gygi et al.).

of identical sequences from the two sources are easily recognized as pairs differing by eight atomic mass units. Differences in their expression levels can be sensitively compared and normalized to correct for differences in overall protein content.

Many other affinity enrichment techniques can be imagined, tailored according to the proteins of interest. For example, lectins can be used to make a column selectively capturing glycoproteins.

13.5 Protein microarrays

Generic aspects of microarrays have already been covered in §13.1. Protein microarrays allow the simultaneous assessment of expression levels for thousands of genes across various treatment conditions and time. The main difference compared with nucleic acid arrays is the difficulty and expense of placing thousands of protein capture agents on the array. Since capture does not depend on simple hybridization, but on a certain arrangement of amino acids in three dimensional space, complete receptor proteins such as antibodies have to be used, and then there is the danger that their conformation is altered by immobilization on the chip surface.[10] It may be possible to exploit the advantages of nucleic acid immobilization (especially the convenient photofabrication method) by using aptamers—oligonucleotides binding specifically to proteins—for protein capture. This approach should at least be useful for microarray experiments to determine the expression levels of transcription factors.

Polypeptide immobilization chemistries typically make use of covalently linking peptide side chain amines or carboxyl groups with appropriately modified chip surfaces. Quite a variety of possible reactions exist but usually several side chains are able to react with the surface, making orientational

---

[10]As an alternative way to prepare receptors, the phage display technique invented by Dyax is very useful. The gene for the coat protein expressed abundantly on the surface of a bacteriophage virus is modified by adding a short sequence coding for an oligopeptide to one end. Typically a large number ($\sim 10^9$) of random oligonucleotides are synthesized and incorporated (one per phage) into the virus gene. The phages are then allowed to multiply by infecting a host virus; the random peptide is expressed in abundance on the coat of the phage along with the regular coat protein. The phage population is then exposed to an immobilized target (e.g. a protein). Any phage (a single one suffices) whose peptide interacts with the target during this screening is retained and is then also expressed in abundance on the coat of the phage. The phage population is then exposed to an immobilized target (e.g. a protein). Any phage (a single one suffices) whose peptide interacts with the target during this screening is retained and recovered, and then multiplied in bacteria.

specificity difficult to achieve. Proteins recombinantly expressed with a terminal oligohistidine chain can be bound to surface-immobilized nickel ions, but the binding is relatively unstable.

A significant problem with protein microarrays is the nonspecific adsorption of proteins. Unfavourably oriented bound proteins, and exposed substrate offer targets for nonspecific adsorption. Pretreatment with a so-called "blocking" protein (seralbumin is a popular choice) is supposed to eliminate the nonspecific adsorption sites, although some interference with specific binding may also result.

As with the transcriptome, statistical analyses of protein microarray data focus on either finding similarity of gene expression profiles (e.g. clustering), or calculating the changes (ratios) between control and treated samples (differential expression).

13.6   Protein expression patterns

Whether the transcriptome or the proteome is measured, the result from each experiment is a list of expressed objects (mRNA or proteins) and their abundances or net rates of synthesis. These abundances are usually normalized so that their sum is unity. Each experiment is therefore represented by a point in protein space (whose dimension is the number of proteins; the distance along each axis is proportional to abundance); each protein is represented by a point in expression space (whose dimension is the number of experiments). The difficulty in making sense of this data is its sheer extent: there are hundreds or thousands of proteins and there may be dozens of experiments (which could, for example, be successive epochs in a growth experiment, or a series of shocks). Hence there is a great need for drastic data reduction.

One approach has already been mentioned (§13.1), viz. to group proteins into blocks whose expression tends to vary in the same way (increase, decrease, remain unchanged). This is the foundation for understanding how genes are linked together into networks, as will be discussed in the next section.

Another approach is to search for global parameters characterizing the proteome. Considering it as "vocabulary" transferring information from genotype to phenotype, it has been found that the distribution of protein abundance follows the same canonical law as the frequency of words in literary texts.[11] The canonical law has two parameters, the informational temperature, which is low for limited expression of the potential gene repertoire,

---

[11]See Vohradský & Ramsden, Ramsden & Vohradský.

and high for extensive expression, and the effective redundancy ρ, which is high when many alternative pathways are active, and low otherwise.

13.7 Regulatory networks

When one or more stimuli arrive at a cell, the affinities of certain proteins for a transcription factor binding site are altered, and mRNA transcription is activated or inhibited, resulting in altered protein abundance, hence altered metabolic activity, hence altered concentrations of metabolites. One wishes to understand the machinery of this process, which can of course be represented as a dynamical system as in §7.3, or as an automaton (§§7.1.1 and 8.2). How can one derive the network from the gene expression data?

In prokaryotes, and possibly some eukaryotes, genes are organized in operons. As already discussed in Chapter 9, an operon comprises a promoter sequence controlling the expression of several genes (positioned successively dowstream from the promoter, whose products may be successive enzymes in a metabolic pathway.[12] In most of the eukaryotes investigated hitherto, a similar, but less clearly delineated arrangement also exists: the same transcription factor may control the expression of several genes, but they may be quite distant from each other along the DNA, indeed even on different chromosomes. Genes observed to be close to each other in expression space are likely to be controlled by the same activator. Each gene can have its own promoter sequence; coexpression is achieved by the transcription factor binding to a multiplicity of sites. Indeed, given that several factors may have to bind simultaneously to the TFBS region in order to modulate expression, control appears to be most commonly of the "many to many" variety, as anticipated many years ago by Wright. Since genes code for proteins which in turn control the expression of other genes, the network is potentially extremely interconnected.

A formal representation follows the lines already adumbrated in Chapters 7 and 8. It is a very useful simplification to consider the model networks to be Boolean, i.e. genes are switched either on or off. To give a flavour of the approach, consider an imaginary little network in which gene A activates the expression of B, B activates A and C, and C inhibits A.[13] This is just an abbreviated way of saying that the translated transcript of A binds to the promoter sequence of B and activates transcription of B, etc. Hence A, B and C form a network, which can be represented by a diagram of immediate

---

[12]Groups of operons controlled by a single transcription factor are called regulons, groups of regulons modulons.

[13]After Vohradský.

effects (cf. figure 8.1), or as a Boolean weight matrix:

$$
\begin{array}{c|ccc}
  & A & B & C \\
\hline
A & 0 & 1 & -1 \\
B & 1 & 0 & 0 \\
c & 0 & 1 & 0
\end{array}
\qquad (13.5)
$$

Reading from top to bottom gives the cybernetic formalization; reading horizontally gives the Boolean rules: A=B NOT C, B=A, C=B. (13.5) can be transformed to produce a stochastic matrix and the evolution of transcription given by a Markov chain. Alternatively the system can be modelled as a neural net in which the evolution of the expression level a_i (i.e. number of copies produced) of the ith protein in time τ is

$$
\tau \frac{da_i}{dt} = \mathcal{F}_i(\sum_j w_{ij} a_j - x_i) - a_i
\qquad (13.6)
$$

where w is an element of the weight matrix (13.5), $\mathcal{F}$ a nonlinear transfer function (e.g. an exponential function), x is an external input (e.g. a delay) and the negative term at the extreme right represents degradation. The Boolean network approach lends itself to elegant, compact descriptions which can easily be extended to hundreds of genes.

A different kind of approach is to represent regulatory networks as gene circuits, in analogy to electrical circuits. The elements of the circuit are included as Boolean logic gates following certain rules as in the example given above.

Chapter 14

Interactions

It is clear from the description of the workings of the living cell (Chapter 9) that the processes of life everywhere depend on interactions between molecules. From the base-pairing of nucleic acids, to the formation of the bilayer lipid membranes enclosing organelles and cells, through to the protein-protein interactions building up supramolecular complexes serving structural ends, or for carrying out reactions, the regulation of gene expression by transcription factors binding to promotors, the operation of the immune system—the list seems to be almost endless—one observes the molecules of life linked together in a web of interactions. The set of all these interactions, i.e. a list of all the molecules, associated with all the other molecules with which some kind of transient association is found, consitutes the interactome.

It has long been realized that the proteins in the cell interact with each other extensively. Indeed, McConkey long ago coined the term 'quinary structure' (of proteins) for this web of interactions. What is new is that there is now a realistic chance of being able to globally characterize the set of protein-protein interactions in a cell.

If the proteins are considered as the nodes of a graph (cf. §7.2), a pair of proteins will be joined by a vertex if the proteins associate with each other. This is in contrast to metabolic networks, in which two metabolites are joined if there is a chemical reaction (catalysed by an enzyme) leading from one to another (§15.4).

On this basis, the 'interactome'—the set of interactions in which a protein could participate—would be characterized by such a graph, or an equivalent list of all the proteins in a cell, each associated with a sublist of the proteins with which they interact.

We note a few preliminary points. Firstly, this graph is potentially extraordinarily large and complex. Even if one confines oneself to the N expressed proteins in a cell, there are $\sim N^2$ potential binary interactions, and vastly more higher order ones. Even if only a small fraction of these interactions actually occur (and some general principles of the stability of dynamical systems, as discussed in Chapter 7, suggest that only about 10% will be), we are still talking about $\sim 10^7$ interactions, assuming about 10^4 expressed proteins (in a eukaryotic cell), and of course 10^8 pairs would have to be screened in order to find the 10%. In a prokaryote, with possibly only 1000 expressed proteins, the situation is better, but still poses a daunting experimental challenge, even without considering that many of those proteins are present in extremely low concentrations.

Secondly, 'interaction' as implied by elementary chemical reactions of the type

$$A + B \overset{k_a}{\rightleftharpoons} C , \qquad (14.1)$$

where C is a complex of A and B, and for which an affinity (or equilibrium) constant K is often loosely defined by

$$K = \frac{ab}{c} \qquad (14.2)$$

where small letters denote concentrations, is nearly always quite inadequate to characterize the association between two proteins. In practical terms, if an experiment is carried out with scant regard to the underlying physical chemistry, even slight differences in the way of carrying out the reaction, or in the way the data is interpreted, could result in considerable differences in the corresponding numerical values attributed to the interaction. At present, the interactome has mostly been assembled on the basis of dichotomous inquiry (i.e. does the protein interact or does it not?) but as technical capabilities improve this is obviously going to change, and it will become important to assign gradations of affinity to the interactions.

Thirdly, the cytoplasm is crowded and compartmentalized. Hence many pairs of proteins potentially able to interact have no chance of encountering each other in practice. Moreover local concentrations of inorganic ions and small molecules, which may greatly influence the strength of an interaction, often differ greatly from place to place within the cell. This gives an advantage to methods probing interactions *in vivo* over those requiring the proteins to be extracted. On the other hand, *in vivo* measurements cannot generally yield data sophisticated enough to go beyond the elementary model of interaction encapsulated by equation (14.1), and mostly cannot go beyond a simple yes/no appraisal of interaction. Besides, unless the *in vivo* technique involves some three dimensional spatial resolution, the result will be an average over different local microenvironments, physiological states etc. Properly designed *in vitro* experiments can reconstitute conditions of a tightly defined physiological and spatially restricted state of a living cell.

Fourthly, it should be emphasized that many protein interactions take place at the internal surfaces of cells, such as lipid membranes. The physical chemistry of the interactome is thus largely the physical chemistry of heterogeneous reactions, not homogeneous ones.

The field can naturally be extended to include the interactions of proteins with other objects, such as DNA, RNA, oligo- and polysaccharides, lipid membranes, etc. Indeed, it is essential to do so in order to obtain a proper representation of the working of a cell. Although the interactome emerged from a consideration of proteins, protein-DNA and protein-saccharide are indeed exceedingly important in the cell. The latter have been given comparatively less attention.[1]

One proposed simplification has been to consider that protein-protein binding takes place via a relatively small number of characteristic polypeptide domains, i.e. a sequence of contiguous amino acids (sometimes referred to as 'modules'). In the language of immunology, a binding module is an epitope. The module concept implies that the interactome could effectively be considerably reduced in size. There is, however, no consistent way of defining the modules. It seems clear that a sequence of contiguous amino acids is inadequate to do so; an approach built upon the dehydron concept[2]

It is useful to consider two types of protein complexes, 'permanent' and 'transient'. By permanent we mean large multiprotein complexes such as the spliceosome (in principle any multisubunit protein) that remain intact during the lifetime of their constituents. On the other hand, transient complexes form and disintegrate constantly as and when required. The interactome is a highly dynamic structure, and this kinetic aspect needs to be included in any complete characterization.

The kinetic mass action law (KMAL) defines K as

$$K = \frac{k_a}{k_d} \tag{14.3}$$

where the ks are the rate coefficients for association (a) and dissociation (d), but as it is a ratio the same value of K results from association reactions that take either milliseconds or years to reach equilibrium. This temporal aspect can have profound influences on the outcome of a complex interaction. Many biological transformations (of the type often referred to as signal transduction) require the sustained presence of A in the vicinity of B in order to effect a change (e.g. of conformation) in B that will then trigger some further event

---

[1]Remarkable specificity is achievable, see e.g. the episematic process described by Popescu & Misevic.

[2]The dehydron (q.v.) is an underwrapped (i.e. under-desolvated) hydrogen bond and is a key determinant of protein affinity.

(e.g. in C, also bound to B). A very well characterized example of this kind of effect is the photolysis of silver halides.[3] Freshly reduced Ag will relax back to Ag^+ if it fails to capture another electron within a characteristic time. This is the origin of low intensity reciprocity failure of photographic film. Similarly, too weak or too brief an exposure of molecule B to molecule A will result in the failure of A to trigger any change in B, hence in C, and so on. Therefore, K alone is inadequate to characterize an interaction.

There are many proteins intermediate between the two extremes of transient and permanent, e.g. transcription factors which must gain a subunit in order to be able to actively bind to a promoter site.

Finally, in these preliminary remarks we recall the evolutionary constraints imposed on change: a mutation enhancing the efficiency of an enzyme may be unacceptable because of adverse changes to its quinary structure.

In the remainder of this chapter we consider the basic types of intermolecular interactions; experimental techniques for determining interactions *in vivo* and *in vitro*; and some notions about the network structure of the interactome, including its dynamical aspects. A special section deals with the immune repertoire, which is a very well studied aspect of whole organism interactions.

14.1 Basic intermolecular interactions

The simplest, and least specific, is hard body exclusion. Atoms cannot interpenetrate due to the Born repulsion. The situation is slightly more complicated for macromolecules of irregular shape, i.e. with protrusions and reëntrant hollows: they may be modelled as spheres with effective radii, in which case some interpenetration may be possible.

The Lifshitz-van der Waals force is nearly always weakly attractive, but since it operates fairly indiscriminately, not only between macromolecules but also between such a molecule and the small solvent molecules, it is of little importance in conferring specificity of interaction.

Most macromolecules are ionized at cytoplasmic pH, due to dissociation (from –COOH) or addition (to –NH_2) of a proton, but the charge is usually effectively screened in the cytoplasmic environment, such that the characteristic distance (the Debye length) of the electrostatic interaction between charged bodies may be reduced to a fraction of a nanometre. Hence it is mainly important for short-range steering prior to docking.

Hydrogen bonds (H-bonds or HB) have already been encountered (§§10.2, 10.3, 10.5 etc.). A chemical group can be either an HB-donor or an HB-

---

[3]See e.g. Ramsden (1984, 1986).

acceptor. Potentiated by water, this interaction can have a considerable range in typical biological milieux—out to tens of nanometres. It is the dominant interparticle interaction in biological systems.[4]

'Hydrophobic effects' 'or 'forces' are also a manifestation of hydrogen bonding in the presence of water, which can effectively compete for intermolecular hydrogen bonds. The wrapping of dehydrons by appropriate apolar residues is a key contributor to protein-protein affinity.

It may be useful to think of the interactions between macromolecules in a cell as analogous to those between people at a party. It is clear that everyone is subject to hard body exclusion. Likewise, one may feel a weak attraction for everyone—misanthropes would not be present. This is sufficient to allow one to fleetingly spend time exchanging a few words with a good many people, among whom there will be a few with strong mutual interest and a longer conversation will ensue. Once such mutual attraction is apparent, the conversation may deepen further, and so on. This is very like the temporal awareness shown by interacting macromolecules, whose probability of remaining together can be described by a memory function: the amount $\nu(t)$ of associated protein can be represented by the integral

$$\nu(t) = k_a \int_0^t \phi(t_1)Q(t,t_1)dt_1 \tag{14.4}$$

where ϕ is the fraction of unoccupied binding sites. The memory kernel Q denotes the fraction of A bound at epoch t_1 which remain adsorbed at epoch t. Often Q simply depends on the difference $t - t_1$. If dissociation is a simple first order (Poisson) process, then $Q(t) = \exp(-k_d t)$ and there is no memory. The dissociation rate coefficient is time dependent and can be obtained from the quotient

$$k_d(t) = \frac{\int_0^t \phi(t_1)Q'(t,t_1)dt_1}{\int_0^t \phi(t_1)Q(t,t_1)dt_1}, \tag{14.5}$$

where Q' is the derivative of the memory function with respect to time. Finally we note that a necessary condition for the system to reach equilibrium is

$$\lim_{t \to \infty} Q(t) = 0. \tag{14.6}$$

Specificity

From the above considerations it follows that specificity of interaction is mainly influenced by geometry (due to hard body exclusion), the pattern

---

[4]Hydrogen bonding is a special example of Lewis acid-base (AB), or electron donor-acceptor (da) interactions.

of complementary arrangements of HB donors and acceptors, (for which an excellent example is the base pairing in DNA and RNA (figures 10.3 and 10.5) and the pattern of complementary arrangements of dehydrons and apolar residues on the two associating partners.

Nonspecific interactions

Most biological interactions show no discontinuity of affinity with some parameter characterizing the identity of one of the binding partners, or their joint identity, although the relation may be nonlinear. Hence in most cases the difference between specific and nonspecific interactions is quantitative, not qualitative. Even nucleotides can pair with the wrong bases, albeit with much smaller affinity. In many cases, such as the association of transcription factors with promoter sites, weak nonspecific binding to any DNA sequence allows early association of the protein with the nucleic acid, whereupon the search for the promoter sequence becomes a random walk in one dimension rather than three, which enormously accelerates the finding process.[5]

Cooperative binding

Consider again the reaction (14.1) with A representing a ligand binding to an unoccupied site on a receptor (B). Suppose that the ligand-receptor complex C has changed properties that allow it to undergo further, previously inaccessible reactions (e.g. binding to a DNA promoter sequence). The rôle of A is to switch B from one of its stable conformational states to another. The approximate equality of the intramolecular, molecule–solvent and A–B binding energies is an essential feature of such biological switching reactions. An equilibrium binding constant K_0 is defined according to the law of mass action (14.2). If there are n independent binding sites per receptor, conservation of mass dictates that $b = nb_0 - c$, where b_0 is the total concentration of B, and the binding ratio $r = c/b_0$ (number of bound ligands per biopolymer) becomes

$$r = \frac{nK_0a}{1 + K_0a} \ . \tag{14.7}$$

Suppose now that the sites are not independent, but that addition of a second (and subsequent) ligand next to a previously bound one (characterized by an equilibrium constant K_1) is easier than the addition of the first ligand. In the case of a linear receptor B, the problem is formally equivalent to the one

---

[5]E.g. Ramsden & Dreier; see Ramsden & Grätzel for a nonbiological example of dimensional reduction from three to two dimensions.

dimensional Ising model of ferromagnetism, and neglecting end effects, one has:

$$r = \frac{n}{2}\left(1 - \frac{1 - K_0 a}{[(1 - K_0 a)^2 + 4K_0 a/q]^{1/2}}\right) \tag{14.8}$$

where the degree of coöperativity q is determined by the ratio of the equilibrium constants, $q = K_1/K_0$. For $q > 1$ this yields a sigmoidal binding isotherm. If $q < 1$, then binding is anticoöperative, as for example when an electrically charged particle adsorbs at an initially neutral surface; the accumulated charge repels subsequent arrivals and makes their incorporation more difficult.

14.2 *In vivo* experimental methods

Several methods have been developed involving manipulations on living cells. Although called *in vivo*, they cannot be called noninvasive. The cell is assaulted quite violently: either it is given unnatural, but not lethal reagents, or it is killed and swiftly analysed before decay sets it, the interactions present at the moment of death being assumed to remain until they have been measured.

14.2.1 The yeast two-hybrid assay

Suppose that it is desired to investigate whether protein A interacts with protein B. The general concept behind this type of assay is to link A to another protein C, and B to a fourth protein D. C and D are chosen such that if they are complexed together (via the association of A and B), they can activate some other process, e.g. gene expression, in yeast. In that case C could be the DNA-binding domain of a transcription factor, and D could trigger the activation of RNA polymerase. The name 'hybrid' refers to the need to make hybrid proteins, i.e. the fusion proteins A-C and B-D. If A indeed associates with B, when A-C binds to the promoter site of the reporter gene, B-D will be recruited and transcription of the reporter gene will begin. The advantage of the technique is that the interaction takes place *in vivo*.

Many variants of the basic approach can be conceived and some have been realized, e.g. A could be anchored to the cell membrane, and D (to which B is fused) could activate some other physiological process if B becomes bound to the membrane.

Disadvantages of the technique include: the cumbersome preparations needed (i.e. making the fusion proteins by genetic engineering); the possible, or even likely, modification of the affinities of A and B for each other, and of

C and D for their native binding partners, through the unnatural fusion protein constructs; and the fact that the interactions take place in the nucleus, which may not be the native environment for the A-B interaction. It is also restrictive that interactions are tested in pairs only (although this does not seem to be a problem in principal; transcription factors requiring three or more proteins to activate transcription could be used.

14.2.2   Crosslinking

The principle of this approach is to instantaneously crosslink all associated partners (protein-protein and protein-DNA) using formaldehyde, while the cell is still alive. It is then lysed to release the crosslinked products, which can be identified by mass spectrometry. In the case of a protein-nucleic acid complex, the protein can be degraded with a protease, and the DNA fragments to which the protein was bound—which should correspond to transcription factor binding sites—can be identified by hybridizing to a DNA microarray.

14.2.3   Correlated expression

The assumption behind this family of methods is that if the responses of two (or more) proteins to some disturbance are correlated, then the proteins are associated.

For example, mRNA expression is measured before and after some change in conditions; proteins showing similar changes in transcriptional response (increase or decrease etc.) are inferred to be associated.

Another approach is to simultaneously delete (knock out) two (or more) genes that individually are not lethal. If the multiple knock out is lethal, then it is inferred that the encoded proteins are associated.

Although these approaches, especially the first, are convenient for screening large numbers of proteins, the assumptions that co-expression implies actual interaction, or that functional association implies actual interaction, are very unlikely to be generally warranted, and indeed strong experimental evidence for it is lacking.

14.2.4   Other methods

Many other ways to identify protein complexes are possible. For example, A could be labelled with a fluorophore, and B with a different fluorophore absorbing and emitting at lower wavelengths. If the cell is illuminated such that A's fluorophore is excited, but the emission of B's fluorophore is observed, then it can be inferred that A and B are in sufficiently close proximity that the excitation energy could be transferred by Förster resonance from one to the other. This approach has a number of undesirable features, such as the need to label the proteins, and the possibility of unfavourable alignment of the fluorophores, such that transfer is hindered even though A and B are associated.

14.3 *In vitro* experimental methods

Here affinities are measured outside the cell. At least one of the proteins of interest has to be isolated and purified. It can then be immobilized on a chromatographic column, and the entire cell contents passed through the column. Any other proteins interacting with the target protein will be bound to the column, and can be identified after elution.

A much more powerful approach, because it allows precise characterization of the kinetics of both association and dissociation, is to immobilize the purified target protein on a transducer able to respond to the presence of proteins binding to the target. The combination of capture layer and transducer is called a biosensor (figure 14.1).

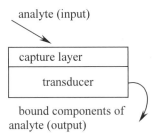

Figure 14.1: Schematic representation of a biosensor. The thickness and structure of the capture layer, which concentrates the analyte whose presence can be registered by the transducer, largely determines the temporal response. The main transducer types are mechanical (cantilevers, the quartz crystal microbalance), electrical (electrodes, field effect transistors), optoelectronic (surface plasmon resonance) and optical (planar waveguides, optical fibres). See Ramsden (1994) and Scheller & Schubert for comprehensive oveviews.

Although this approach is formally *in vitro*, the physiological milieu can be reproduced to practically any level of detail. Indeed, as pointed out in the previous section, the microenvironment of a subcellular compartment can be more precisely investigated than *in vivo*. Nevertheless, since each interaction is individually measured, with as much detail as is required, high throughput is only possible with massive parallelization, but because of the current expense of transducing devices, this parallelization is only practical with protein microarrays, the penalty of which is that almost all kinetic information is lost. Hence at present, protein microarrays and serial direct affinity measurement using biosensing devices are complementary to each other. Miniaturization of the transducers and large scale integration of arrays of devices (comparable to the development of integrated circuit technology from individual transistors, or the development of displays in which each pixel is driven by a tiny circuit behind it) will allow the essential detailed kinetic characterization to be carried out in a massively parallel mode. Significant improvements in microarrays, allowing reliable kinetic information to be obtained from them, are also envisaged. In effect, the two approaches will converge.

14.3.1   Chromatography

Chromatography denotes an arrangement whereby one binding partner is immobilized to a solid support (the stationary phase), and the other partner is dissolved or dispersed is a liquid flowing past the solid (the mobile phase). In essence it is like the biosensor: the difference is that binding is not measured *in situ*, but by depletion of the concentration of the mobile in the output stream. As with the biosensor, a drawback is that the immobilized protein has to be chemically modified in order to be bound to the immobile phase of the separation system. In contrast to the biosensor, the hydrodynamics within the column are complicated and chromatography is not very useful for investigating the kinetics of binding. On the other hand there is usually an immense area of surface within the column, and the technique is therefore useful for preparative purposes.

Typically the protein complexes are identified using mass spectrometry (examples of methods are tandem affinity purification, TAP, or high-throughput mass spectrometric protein complex identification, HMS-PCI).

14.3.2   Direct affinity measurement

As indicated in figure 14.1, a variety of transducers exist, the most popular being the quartz crystal microbalance (QCM), surface plasmon resonance

(SPR), and optical waveguide lightmode spectroscopy (OWLS).[6] A great advantage of biosensors is that no labelling of the interacting proteins is required, since the transducers are highly sensitive. The order of intrinsic sensitivity is QCM < SPR < OWLS. The most sensitive method (i.e. OWLS) can easily detect one protein per 50 μm^2.

QCM and SPR present a metal surface to the recreated cytoplasm, to which it can be problematical to immobilize one of the binding partners.[7] OWLS has no such restriction since the transducer surface can be any high refractive index transparent material. Titania is a popular material. Moreover the risk of denaturing the protein by the immobilization procedure can be avoided by coating the transducer (the optical waveguide) with a natural bilayer lipid membrane, and choosing a membrane-associated protein as the target.

For measuring the interaction, one simply causes a solution of the putative binding protein (A) to flow over its presumed partner (B) immobilized at the transducer surface. the binding of A to B can be recorded with very high time resolution. The method has already been extended to molecules other than proteins.

The real power of this approach lies in the comprehensive characterization (i.e. precise determination of the number of associated proteins with good time resolution) of the association that it can deliver. A major defect of the description built around equation (14.1) is that the dissociation of A from B is only very rarely correctly given by an equation of the type $d\nu/dt \sim e^{-k_d t}$ (where ν is the number of associated proteins), i.e. a pure Poisson process without memory, since most proteins remember how long they have been associated. This is consequence of the fact that they have several stable states and transitions between the states can be induced by a change in external conditions, such as binding to another protein. The correct approach is to consider that during a small interval of time Δt_1 at epoch t_1, a number $\Delta \nu$

---

[6]See Ramsden (1994) for a comprehensive survey of all these and others.

[7]A popular way to avoid the bioincompatibility of the gold or silver surface of the transducer required with SPR has been to coat it with a thick ($\sim$ 200 nm) layer of a biocompatible polysaccharide such as dextran, which forms a hydrogel, to which the target protein is bound. Unfortunately this drastically changes the transport properties of the solution in the vicinity of the target (bound) protein (see the paper by Schuck), which can lead to errors of up to several orders of magnitude in apparent binding constants (via a differential effect on k_a and k_d). Furthermore such materials interact very strongly (via hydrogen bonds) with water, altering its hydrophilicity, with concomitant drastic changes to protein affinity, leading to further, possibly equally large, distortions in binding constant via its link to the free energy of interaction, i.e. $\Delta G = -RT \ln K$.

of molecules of A will be bound to B, hence

$$\Delta\nu = k_a(\nu, t_1)\, c_A(\nu, t_1)\, \phi(\nu, t_1)\, \Delta t_1 \qquad\qquad (14.9)$$

where c_A is the concentration of free (unassociated) A and ϕ the probability that there is room to bind (we recall that the cell is a very crowded milieu). The memory function $Q(t, t_1)$ gives the probability that a molecule bound at epoch t_1 is still bound at a later epoch t, hence (cf. equation 14.4)

$$\nu(t) = \int_0^t k_a(t_1)c_A(t_1)\phi(t_1)Q(t, t_1)\mathrm{d}t_1 \ . \qquad\qquad (14.10)$$

The memory function, as well as all the other parameters in (14.10), can be found from the high resolution association and dissociation kinetics.

Further advantages of the biosensor approach include the ability to study collective and cooperative effects, and to determine the precise stoichiometry of the association.

14.3.3   Protein chips

In order to enable many interactions to be measured simultaneously, microarrays have been developed. They are identical in concept to nucleic acid microarrays (see §13.5, except that instead of nucleic acids either deposited or synthesized *in situ*, proteins are deposited. The immobilization of proteins without altering their conformation, and hence association characteristics, is much more difficult than for nucleic acid oligomers, however. With these arrays, the interaction of protein A with thousands of other proteins can be studied in a single experiment, by letting A flow over the array. At present, some kind of marking of A (e.g. post-reaction staining) is required to allow its presence at certain sites on the array to be identified. The physical chemistry of operation of these devices is governed by the same basic set of equations as for the biosensor approach, although it is not presently possible to achieve the same sensitivity and time resolution.

14.4   Interactions from sequence

The principle of this approach is that gene proximity is the result of selective evolutionary pressure to associate genes that are co-regulated and, hence, possibly interacting. The motivation is to develop a method that is far less tedious and labour intensive (and hence expensive) than the experimental techniques discussed in the preceding two sections, yet no less accurate (or not more inaccurate).

Certain proteins (in a single given species) apparently consist of fused domains corresponding to individual proteins (called component proteins) in other species. The premiss of the method is that if a composite (fused or fusion) protein in one species is uniquely similiar to two component proteins in another species, which may not necessarily be encoded by adjacent genes, those component proteins are likely to interact. 'Interaction' may be either physical association or indirect functional association such as involvement in the same biochemical pathway, or co-regulation. Hence what is inferred from this method does not exactly correspond with what is measured in the experimental methods. Nevertheless is is an interesting attempt, and one which could be developed with more sophistication, to extract interaction data from sequence alone, which is a kind of Holy Graal for interactomics, since it is so much easier nowadays to obtain sequence data than any other kind.

14.5   The immune repertoire

Antibodies binding to an antigen interact with a relatively small portion of the molecule. The number N of foreign antigens which must be recognized by an organism is very large, perhaps greater than 10^{16}, and at the same time there is a smaller number, $N' \sim 10^6$, of self-antigens which must *not* be recognized. Yet the immunoglobulin and T-cell receptors may only contain $n \sim 10^7$ different motifs. Recognition is presumed to be accomplished by a generalized lock and key mechanism involving complementary amino acid sequences. How large should the complementary region be, supposing that the system has evolved to optimize the task?[8] If P_S is the probability that a random receptor recognizes a random antigen, the value of its complement $P_F = 1 - P_S$ maximizing the product of the probabilities that each antigen is recognized by at least one receptor, and that none of the self-antigens is recognized, i.e. $(1 - P_F^n)^N P_F^{nN'}$, is:[9]

$$P_F = \left(1 + \frac{N}{N'}\right)^{-1/n} . \tag{14.11}$$

Using the above estimates for n, N and N', one computes $P_S \approx 2 \times 10^{-6}$. Suppose that the complementary sequence is composed of m classes of amino acids and that at least c complementary pairs on a sequence of s amino acids are required for recognition. Since the probability of a long match is very

---

[8]A similar problem is posed by the olfactory system.
[9]Percus et al.

small, to a good approximation the individual contributions to the match can be regarded as being independent. A pair is thus matched with probability $1/m$, and mismatched with probability $1 - 1/m$. Starting at one end of the sequence, runs of c matches occur with probability m^{-c}, and elsewhere they are preceded by a mismatch and can start at $s - c$ possible sites. Hence

$$P_S = [(s - c)(m - 1)/m + 1]/m^c . \tag{14.12}$$

If $s \gg c > 1$, one obtains

$$c = \log_m[s(m - 1)/m] - \log_m P_S . \tag{14.13}$$

Supposing s to be a few tens, $m = 3$ (positive, negative and neutral residues), and again using the numbers given above (since they all enter as logarithms the exact values are not critical) one estimates $c \sim 15$, which seems to be in good agreement with observation.

Effective stimulation in the immune system often depends on a sustained surface reaction. When a ligand (antigen) present at the surface of an antigen presenting cell (APC) is bound by a T lymphocyte (TL), binding triggers a conformational change in the receptor protein to which the antigen is fixed, which initiates further processes within the APC, resulting in the synthesis of more receptors, etc. This sustained activation can be accomplished with a few, or even only one TL, provided that the affinity is not too high: the TL binds, triggers one receptor, then dissociates and binds anew to a nearby untriggered receptor (successive binding attempts in solution are highly correlated). This 'serial triggering' can formally be described by:

$$L + R \rightarrow R_L^* \tag{14.14}$$

(with rate coefficient k_a) where the starred R denotes an activated receptor, and

$$R_L^* \rightleftharpoons R^* + L \tag{14.15}$$

with rate coefficient k_d for dissociation of the ligand L from the activated receptor, and the same rate coefficient k_a for reassociation of the ligand with an already activated receptor. The rate of activation (triggering) is $-dr/dt = -k_a rl$, solvable by noting that $d\,l/dt = -k_a(r + r^*) + k_d r_L^*$. One obtains

$$l(t) = \frac{k_a \tau}{1 - Y e^{-t/\tau}} + \frac{k_a(l_0 - r_0) - k_d - 1/\tau}{2k_a} \tag{14.16}$$

where $\tau = \{4l_0 k_a k_d + [k_a(l_0 - r_0) - k_d]^2\}^{-1/2}$ and $Y = (k_d + k_a[l_0 + r_0] - 1/\tau)/(k_d + k_a[l_0 + r_0] + 1/\tau)$, subscripts 0 denoting the initial concentrations of R and L, and the sought-for solution is then

$$r(t) = r_0 \exp\left[\ln\left(\frac{1 - Y e^{-t/\tau}}{1 - Y}\right) - \frac{t}{\tau}\right] . \tag{14.17}$$

14.6 Small systems

Biological reactions, especially those *in vivo*, typically take place in very confined volumes. This may have a profound effect on the kinetic mass action law (KMAL). Consider the reaction $A + B \xrightarrow{k_a} C$, which Rényi (1953) has analysed in detail. We have

$$\frac{dc}{dt} = k_a[\bar{a}\bar{b} + \Delta^2(\gamma_t)] = k_a\overline{ab} \qquad (14.18)$$

where the bars denote expected numbers of A and B, and γ_t is the number of C molecules created up to time t. The term $\Delta^2(\gamma_t)$ expresses the fluctuations in γ_t: $\overline{\gamma_t^2} = \overline{\gamma_t}^2 + \Delta^2(\gamma_t)$: supposing that γ_t approximates to a Poisson distribution, then $\Delta^2(\gamma_t)$ will be of the same order of magnitude as $\overline{\gamma_t}$. The KMAL, which puts $\bar{a} = a_0 - c(t)$ etc., the subscript 0 denoting initial concentration (at $t = 0$), is a first approximation in which $\Delta^2(\gamma_t)$ is supposed negligibly small compared to $\bar{a}$ and $\bar{b}$, implying that $\bar{a}\bar{b} = \overline{ab}$, whereas strictly speaking it is not since a and b are not independent: the disappearance of A at a certain spot (i.e. its transformation into C) implies the simultaneous disappearance of B. The neglect of $\Delta^2(\gamma_t)$ is justified for molar quantities of starting reagents,[10], but not for reactions in minute subcellular compartments. The number fluctuations, i.e. the $\Delta^2(\gamma_t)$ term, will constantly tend to be eliminated by diffusion. This generally dominates in macroscopic systems. When diffusion is hindered however, because of the correlation between a and b, initial inhomogeneities in their spatial densities lead to the development of zones enriched in either one or the other faster than the enrichment can be eliminated by diffusion. These consequences of small systems place fundamental limitations on biological processes such as gene regulation.

14.7 Global statistics of interactions

The experimental difficulties are still so onerous, the uncertainties so great, and the amount of data so little that researchers have mostly been content to draw diagrams, essentially graphs, of their results, with the proteins as nodes and the associations as vertices, and leave it at that; at most a difference in the pattern between a pair of sets of results from the same organism grown under two different conditions might be attempted. An amusing endeavour to go beyond this first stage of representation has been made,[11] with the result

---

[10]Except near the end of the process, when $\bar{a}$ and $\bar{b}$ become very small.

[11]Jeong et al.

(from a single dataset covering protein-protein interactions in yeast, with just under 1900 proteins and just over 2200 interactions) that the probability that a given protein interacts with k other proteins follows a power law over about one and a half orders of magnitude with an exponent ~ -2. Unsurprisingly, the most heavily connected proteins were also found to be the most likely to cause lethality if knocked out.

Networks of interactions are adjunct to the gene networks considered in the previous chapter, and are mainly used to assist in compiling the gene interaction network of an organism, since in general the interaction between two biomacromolecules only has sense in a regulatory context.

Chapter 15

Metabolomics and metabonomics

Metabolism is the ensemble of chemical transformations carried out in living tissue; operationally it is embodied in the matter and energy fluxes through organisms. Metabolomics is defined as the measurement of the amounts (concentrations) and locations of the all the metabolites in a cell, the metabolites being the small molecules ($M_r \lesssim 1000$, e.g. glucose, cAMP,[1] GMP,[2] glutamate etc.) transformed in the process of metabolism, i.e. mostly the substrates and products of enzymes.[3] The quantification of the amounts of expressed enzymes is, as we have seen, proteomics; metabolomics is essentially an extension of proteomics to the activities of the expressed enzymes, and it is of major interest to examine correlations between expression data and metabolite data.[4]

Metabonomics is a subset of metabolomics and is defined as the quantitative measurement of the multiparametric metabolic responses of living systems to pathophysiological stimuli or genetic modification, with particular emphasis on the elucidation of differences in population groups due to

---

[1] Cyclic adenosine monophosphate.

[2] Guanosine monophosphate.

[3] The "official" classification of enzyme function is that of the Enzyme Commission (EC), which recognizes six main classes: 1, oxidoreductases; 2, transferases; 3, hydrolases; 4, lyases; 5, isomerases; and 6, ligases. The main class number is followed by three further numbers (separated by points), whose significance depends on the main class. For class 1, the second number denotes the substrate and the third number the acceptor; whereas for class 3, the second number denotes the type of bond cleaved and the third number the molecule in which that bond is embedded. For all classes, the fourth number signifies some specific feature such as a particular cofactor.

[4] These correlations are crucial for understanding the links between genome and epigenetics.

genetic modification, disease and environmental stress. In the numerous cases of diseases not obviously linked to genetic alteration (mutation), metabolites are the most revealing markers of disease or chronic expose to toxins from the environment, and of the effect of drugs. As far as drugs are concerned, metabonomics is effectively a subset of the investigation of the absorption, distribution, metabolism and excretion (ADME) of drugs.

Metabonomics usually includes not only intracellular molecules but also the components of extracellular biofluids. Of course many such molecules have been analysed in clinical practice for centuries; the novelty of metabonomics lies above all in the vast increase of the scale of analysis; high throughput techniques allow large numbers (hundreds) of metabolites to be analysed simultaneously and repeat measurements can be carried out in rapid succession, enabling the temporal evolution of physiological states to be monitored. The concentrations of a fairly small number of metabolites has been shown in many cases to be so well correlated with a pathological state of the organism that these metabolite concentrations could well serve as the essential variables of the organism, whose physiology is, as we may recall, primarily directed towards maintaining the essential variables within viable limits.

Metabonomics is being integrated with genomics and proteomics in order to create a new systems biology, fully cognizant of the intense interrelationships of genome, proteome and metabolome. For example, ingestion of a toxin may trigger expression of a certain gene, which is enzymatically involved in a metabolic pathway, thereby changing it, and those changes may in turn influence other proteins, and hence (if some of those proteins are transcription factors or cofactors) gene expression.

15.1   Data collection

The basic principle is the same as in genomics and proteomics: separation of the components followed by their identification. Unlike genomics and transcriptomics, metabonomics has to deal with a diverse set of metabolites even more varied than proteins (which are at least all polypeptides). Typical approaches are to use gas chromatography to separate the components one is interested in, and mass spectrometry to identify them. Alternatively, high resolution nuclear magnetic resonance spectroscopy can be applied directly to many biofluids and even organ or tissue samples

Metabolic microarrays operate on the same principle as other kinds of microarrays (§13.1) in which large numbers of small molecules are synthesized, typically using combinatorial or other chemistry for generating high diversity. The array is then exposed to the target, whose components of interest are usually labelled. This technique can be used to answer questions such as "to which metabolite(s) does macromolecule X bind?"

Much ingenuity is currently being applied to determine spatial variations in selected metabolites. An example of a new method developed for that purpose is PEBBLES (Probes Encapsulated By Biologically Localized Embedding): fluorescent dyes, entrapped inside larger cage molecules, and which respond (i.e. change their fluorescence) to certain ions or molecules. Their spatial location can be mapped using fluorescence microscopy. Another example is the development of high resolution scanning secondary ion mass spectrometry ("nanoSIMS"), whereby a focused ion beam (usually Cs^+ or O^-) is scanned across a (somewhat conducting) sample and the secondary ions released from the sample are detected mass spectrometrically with a spatial resolution of some tens of nanometres. This method is very favourable for certain metal ions, which can be detected at mole fractions of as little as 10^{-6}. If biomolecules are to be detected, it is advantageous to label the molecule or molecules of interest with non-natural isotopes, e.g. ^{15}N; the enriched molecule can then easily be distinguished via the masses of its fragments in the mass spectrometer.

15.2 Data analysis

The first task in metabonomics is typically to correlate the presence of metabolites with gene expression. One is therefore trying to correlate two datasets, each containing hundreds of points, with each other. This in essence a problem of pattern recognition. There are two categories of algorithms used for this task, unsupervised and supervised.

The unsupervised techniques determine whether there is any intrinsic clustering within the dataset. Initial information is given as object descriptions but the classes to which the objects belong is not known beforehand. A widely used unsupervised technique is principal component analysis (PCA). Essentially the original dataset is projected onto a space of lower dimension. For example, a set of metabonomic data consisting of a snapshot of the concentrations of one hundred metabolites is a point in a space of one hundred dimensions. One rotates the original axes to find a new axis along which there is the highest variation in the data. This axis becomes the first principal component. The second one is orthogonal to the first, and has the highest residual variation (i.e. that remaining after the variation along the first axis has been taken out), the third axis is again orthogonal, and has the next highest residual variation, and so on. Very often the first two or three axes are sufficient to account for an overwhelming proportion of the variation in the original data. Since they are orthogonal, the principle components are uncorrelated (zero covariance).

In supervised methods, the initial information is given as learning descriptions, i.e. sequences of parameter values (features) characterizing the object whose class is known beforehand.[5] The classes are non-overlapping. During the first stage, decision functions are elaborated enabling new objects from a dataset to be recognized, and during the second stage, those objects are recognized. Neural networks are often used as supervised methods.

15.3 Metabolic regulation

Once all the data has been gathered and analysed, one attempts to interpret the regularities (patterns). *Simple regulation* describes the direct chemical relationship between regulatory effector molecules, together with their immediate effects, such as feedback inhibition of enzyme activity or the repression of enzyme biosynthesis. *Complex regulation* deals with specific metabolic symbols and their domains. These 'symbols' are intracellular effector molecules that accumulate whenever the cell is exposed to a particular environment. Their domains are the metabolic processes controlled by them. For example, hormones encode a certain metabolic state; they are synthesized and secreted, circulate in the blood, and finally are decoded into primary intracellular symbols (§15.3.2).

15.3.1 Metabolic control analysis (MCA)

This is the application of systems theory (§7.1) or synergetics (§7.3) to metabolism. Let $\mathbf{X} = \{x_1, x_2, \ldots, x_m\}$, where x_i is the concentration of the ith metabolite in the cell, i.e. the set $\mathbf{X}$ constitutes the metabolome. These concentrations vary in both time and space. Let $\mathbf{v} = \{v_1, v_2, \ldots, v_r\}$, where v_j is the rate of the jth process. To a first approximation, each process corresponds to an enzyme. Then

$$\frac{d\mathbf{X}}{dt} = \mathbf{Nv} \tag{15.1}$$

where the "stoichiometry matrix" $\mathbf{N}$ specifies how each process depends on the metabolites. Metabolic control theory (MCT) is concerned with solutions to equation (15.1) and their properties. The dynamical system is generally too complicated for explicit solutions to be attempted, and numerical solutions are of little use unless one knows more of the parameters (enzyme rate

---

[5]See e.g. Tkemaladze.

coefficients) and can measure more of the variables than are generally available at present. Hence much current discussion about metabolism centres on qualitative features. Some are especially noteworthy: it is well known, from numerous documented examples, that large changes in enzyme concentration may cause negligible changes in flux through pathways of which they are a part. Metabolic networks are truly many-component systems, as discussed in Chapter 7, and hence the concept of feedback, so valuable in dealing with systems of just two components, is of little value in understanding metabolic networks.

Problem. Write **X** and **v** in equation 15.1 as column matrices, and **N** as an $m \times r$ matrix. Construct and solve an explicit example with only two or three metabolites and processes.

15.3.2 The metabolic code

It is apparent that certain molecules mediating intracellular function, e.g. cAMP, are ubiquitous in the cell (see table 15.1). Tomkins has pointed out that these molecules are essentially symbols encoding environmental conditions. The domain of these symbols is defined as the metabolic responses controlled by them. Note that symbols are metabolically labile, and are not chemically related to molecules promoting their accumulation. The concept applies to both within and without cells. Cells affected by a symbol may secrete a hormone, which circulates (e.g. via the blood) to another cell, where the hormone-signal is decoded—often back into the same symbol.

Condition	symbol	domain
glucose deficiency	cAMP	starvation response
N-deficiency	ppGpp	stringent response
redox level	NADH	DNA transcription

Table 15.1: Some examples of metabolic coding.

15.4 Networks

Metabolism can be represented as a network in which the nodes are the enzymes and the edges connecting them are the substrates and products of the enzymes. There are two main lines of investigation in this area, which have hitherto been pursued fairly independently from one another.

The first approach is centred on metabolic pathways, defined as series of consecutive enzyme-catalysed reactions producing specific products; 'intermediates' in the series are defined as substances with a sole reaction producing them and a sole reaction consuming them. The complexity of the ensemble of metabolic pathways in a cell is typified by Gerhard Michal's famous chart found on the walls of biochemistry laboratories throughout the world. Current work focuses on ways of rendering this ensemble tractable. For example, a set of transformations can be decomposed into elementary flux modes. An *elementary flux mode* is a minimal set of enzymes able to operate at steady state for a selected group of transformations ('minimal' implies that inhibition of any one enzyme in the set would block the flux). A related approach is to construct linearly independent basis vectors in flux space, combinations of which express observed flux distributions. The extent to which the requirement of a steady state is realistic for living cells remains an open question.

The second approach is to disregard the dynamic aspects and focus on the distribution of the density of connexions between the nodes. The number of nodes of degree k appears to follow a power law distribution, i.e. the probability that a node has k edges $\sim k^{-\gamma}$.[6] There is moreover evidence that metabolic networks thus defined have small world properties (cf. §7.2).

Just as in the abstract networks (automata) discussed previously (Chapter 7), a major challenge in metabolomics is to understand the relationship between the physical structure (the nodes and their connecting edges), and the state structure. As the elementary demonstrations showed (cf. the discussion around figure 7.1), physical and state structures are only tenuously related. Much work is still needed to integrate the two approaches to metabolic networks, and further integrate metabolic networks into expression networks. Life is represented by essentially one network, in which the nodes are characterized by both their amounts and their activities, and the edges likewise.

---

[6]See Wagner & Fell or Raine & Norris.

Chapter 16

Medical applications

Much of the business of bioinformatics concerns the correlation of phenotype with genotype, with the transcriptome and proteome acting as intermediaries.[1] Bioinformatics gives an unprecedented ability to scrutinize the intermediate levels and establish correlations far more extensively and in far more detail than was ever possible before. This ability is revolutionizing medicine. One may represent the human being as a gigantic table of correlations, comprising successive columns of genes and genetic variation, protein levels, and physiological states and interactions.[2]

Medicine is mainly concerned with investigating physiological disorders, and the techniques of bioinformatics allows one to establish correlations between those disorders and variations in the genome and proteome of a patient. Medical applications are mainly concerned with deleterious genetic variation, and with abnormal expression patterns. One can also include drug discovery as a medical application.

---

[1] Indeed one could view the organism as a gigantic hidden Markov model (§12.5.3), in which the gene controls switching between physiological states via protein expression. Unlike the simpler models considered before, here the outputs could intervene in the hidden layers.

[2] Since the physiological column includes entries for neurophysiological states, it might be tempting to continue the table by adding a column for the conscious experiences corresponding to the physiological and other entries. One must be careful to note, however, that conscious experience is in a different category from the columns which precede it. Hence correlation cannot be taken to imply identity (in the same way, a quadratic equation with two roots derived by a piece of electronic hardware is embodied in the hardware, but it makes no sense to say that the hardware has two roots, despite the fact that those roots have well-defined correlates in the electronic states of the circuit components).

16.1   The genetic basis of disease

Some diseases have a clear genetic signature, e.g. normal individuals have about 30 repeats of the nucleotide triplet CGG, whereas patients suffering from fragile X syndrome have hundreds or thousands.

More and more data on the genotype of individuals is being gathered. Millions of single nucleotide polymorphisms (SNPs) are now documented, and studies involving the genotyping of hundreds of SNPs in thousands of people are now feasible. As pointed out earlier, most of the genetic variability across human populations can be accounted for by SNPs, and most of the SNP variation can be grouped into a small number of haplotypes. This growing database is extremely useful for elucidating the genetic basis of disease, or susceptibility to disease, and hence preventive treatment for those screened routinely.

The wish to develop preventive screening implies a need for a much more rapid, and cheaper, way of screening for mutations than is possible with genome sequencing. The classic method is to digest the gene with restriction enzymes and analyse the fragments separated chromatographically using Southern blotting. Although direct genotyping with allele-specific hybridization is possible in simple genomes (e.g. yeast), the complexity of the human genome renders this approach less reliable. Microarrays are extensively applied to this task, as well as a related approach in which the oligonucleotides are attached to small microspheres (beads) a few micrometres in diameter. In effect each bead corresponds to one spot on a microarray. The beads are indivudually tagged, e.g. using a combination of a small number of different attached fluorophores, or via the ratio of two fluorophores. Several hundred different types of beads can be mixed and discriminated at the current level of the technology. A major difficulty in the use of binding assays (hybridization) based on gene chips or beads for allele detection is the lack of complete discrimination between completely matched and slightly mismatched sequences. An alternative approach is based on the very high sequence specificity of certain enzyme reactions, such as restriction.

As well as trying to identify genes, or gene variants, reponsible for disease by analysing the genome of patients, gene segments can be cloned into cells and examined for disease-like symptoms (including the pattern of expression of certain proteins). This approach is called functional cloning.

Much effort goes into understanding the correlation between gene association and disease. There are doubtless many diseases enabled by combinations of two or more variant genes. The problem of correlation then acquires a combinatorial aspect and it becomes much more difficult to solve. The rôle of genomics in medicine will expand once gene syntax is understood. Not least because of the vast amounts of data which have to be dealt with is information science and technology indispensible to the field.

Forensic medicine is an important branch of the medical application of genetic analysis. Repeated motifs such as variable number of tandem repeats (VNTRs), or short tandem repeats (STRs) appear to be uniquely different for each individual. Degradation of the DNA samples, which may have been exposed to adverse environmental influence before collection, limits the use of the longer VNTRs. The smaller STRs require PCR amplification in order to ensure that enough material is available for detection after chromatographic separation. Similar techniques are used to identify microorganisms used in biological warfare and their origin.

16.2 Diagnosis

Many diseases have no clear genetic signature, or depend in a complex way on genetic sequence. In cancer, for example, the relationship between gene and disease must be highly complex, and has so far eluded discovery. Mutations are important (see table 16.1) but the changes in protein levels are equally striking. Both gene and protein chips are important here.[3]

Stage	macro-description	micro-description
A	de-differentiated tissue (atavism)	inherited mutations
B	benign epithelial cancer	acquired mutations: more exposed to carcinogens from the environment
C	adeno-carcinoma	*p53* gene involved
D	metastasis	many mutations

Table 16.1: Stages of a cancer and their genetic correlates.

Knowledge of protein expression patterns greatly expands the knowledge of disease at the molecular level. The full power of the pattern recognition

---

[3]An example of the lack of a simple genetic cause of disease is illustrated by the fact that the same mutations affecting the calcium channel protein in nerve cells are observed in patients whose symptoms range from sporadic headaches to partial paralysis lasting several weeks. This is further evidence in favour of Wright's 'many gene, many enzyme' hypothesis as opposed to Beadle and Tatum's 'one gene, one enzyme' idea.

techniques discussed earlier can be brought to bear in order to elucidate the hidden mechanisms of physiological disorder. The technology of large scale gene expression allows one to correlate gene expression patterns with disease symptoms. Microarray technology has the potential for enabling swift and comprehensive monitoring of the expression profile of a patient. Where correlations become well-established through the accumulation of vast amounts of data, the expression profile becomes useful for diagnosis, and even for preventive treatment of a condition enhancing susceptibility to infection or allergy. One does not simply seek to correlate the bald list of expressed proteins and their abundances with disease symptoms, however: the subtleties of network structure and gene circuit topology, etc. are likely to prove more revealing as possible "causes".

The differential expression of genes in healthy and diseased tissue is usually highly revealing. For the purposes of diagnosis, each gene is characterized as a point in two dimensional space, the two coordinates corresponding to the relative abundance of the gene product in the healthy and diseased tissue. This allows rapid visual appraisal of expression differences.

The composition of blood is also a highly revealing diagnostic source. As well as intact peptides and other biomacromolecules, fragments of larger molecules may also be present. For their identification, mass spectrometry seems to be more immediately applicable than microarrays.

Gene chips also allow the clear and unambigous identification of foreign DNA in a patient due to an invading microorganism, obviating the laborious work of attempting to grow the organism in culture and then identify it phenotypically.

16.3   Drug discovery and testing

Whereas traditionally drugs were sought which bound to enzymes, blocking their activity, bioinformatics-driven drug discovery focuses on control points.

Intervention using drugs can take place very effectively at control points, as summarized in table 16.2. The results of expression experiments are thus carefully scrutinized in order to identify possible control points. One a gene or set of genes have been found to be associated with a disease, they can be cloned into cells and the encoded protein or proteins can be investigated as in more detail as drug targets (functional cloning).

The proteome varies between tissues, and many different structural forms of a protein can be made by a given gene depending on cellular context and the impact of the environment on that cell. From the viewpoint of drug discovery, there are further crucial levels of detail that need to be considered, namely the way that proteins subdivide structurally into discrete domains and how these domains contain small cavities (active sites) which are considered to be the 'true' targets for small-molecule drugs.

Stage	control (examples)
genome → transcriptome (transcription)	epigenetic regulation (networks)
transcriptome → proteome (translation)	posttranslational modification
proteome → dynamic system	distributed control networks
dynamic system → phenotype	hormones

Table 16.2: Stages of gene expression and their control.

Clustering and other pattern recognition techniques discussed earlier are used to identify control points in regulatory networks from proteomics and metabolomics data. DNA, RNA and proteins are thus the significant biological entities with respect to drug development. The stages of drug development are summarized in table 16.3. Great effort is being put into short-cutting this lengthy (and very expensive) process. For example, structural genomics can be used to predict the three dimensional structure of a protein suspected to be at a control point from the corresponding gene sequence. It may also be possible comparison), or at least the active site, or 'specificity pockets' (these regions are typically highly conserved). Toxicogenomics refers to the use of microarrays to evaluate the (adverse) effects of drugs (and toxic substances generally) across a wide range of genes, and pharmacogenomics to the genotyping of patients in an attempt to correlate genotype and response to a drug.

Proteins in cells do not exist in isolation. They bind to other proteins in to form multiprotein structures that *inter alia* are the elements of pathways that control functions such as the response to hormones, allergens, growth signals and so on—things that go wrong in disease. Knowledge of the network of interactions is needed to understand which proteins are the best drug targets. One hopes to develop a physical map of the cell that will allow interpretation of masses of data through mining techniques, and will help train predictive methods for calculating pathways and how they mesh together. Then by honing in on atomic detail of active sites the best candidate drug targets—probably a very small proportion of biologically-valid targets—can

Stage	outcome	technologies involved
1. Target selection	a gene	(functional) genomics; genotyping
2. Protein expression	a 3D protein structure	protein chemistry
3. Screening	a drug which binds	binding studies
4. ADME	a usable drug	interaction studies
5. Trials	an efficacious drug	clinical trials

Table 16.3: Stages of drug discovery and development.

be identified and subjected to closer scrutiny.

Bibliography

This bibliography lists both works on specific topics whose authors are mentioned in the text and sources consulted and felt to be valuable for further reading, but not specifically cited; their utility should be apparent from the title of the book or article.

C. Adami & N.J. Cerf, Physical complexity of symbolic sequences. Physica D 137 (2000) 62–69.

M. Ageno, Linee di ricerca in fisica biologica. Accad. naz. Lincei 102 (1967) 3–50.

M. Ageno, *La "Macchina" Batterica*. Rome: Lombardo Editore (1992).

P.M. Allen, Evolving complexity in social science. In: *Systems—New Paradigms for the Human Sciences* (eds G. Altman & W.A. Koch). Berlin: Walter de Gruyter (1998).

L. Allison, D. Powell & T.I. Dix, Compression and approximate matching. Computer J. 42 (1999) 1–10.

W. Arber, Molecular mechanisms of biological evolution. In: *Frontiers in Biology* (eds C.-H. Chou & K.-T. Shao), pp. 19–24. Taipei: Academia Sinica (1998).

R.B. Ash, *Information Theory*. New York: Interscience (1965).

R.B. Ash, *A Primer of Abstract Mathematics*. Washington (D.C.): Mathematical Association of America (1998).

W.R. Ashby, *An Introduction to Cybernetics*. London: Chapman and Hall (1956).

B. Audit et al., Long-range correlations between DNA bending sites: relation to the structure and dynamics of nucleosomes. J. mol. Biol. 316 (2002) 903–918.

P. Bak & K. Sneppen, Punctuated equilibrium and criticality in a simple model of evolution. Phys. Rev. Lett. 71 (1993) 4083–4086.

E.N. Baker & R.E. Hubbard, Hydrogen bonding in globular proteins. Prog. Biophys. mol. Biol. 44 (1984) 97–179.

A.D. Baxevanis & B.F.F. Ouellette (eds) *Bioinformatics*, 2nd edn. New York: Wiley (2001).

C.H. Bennett, The thermodynamics of computation—a review. Int. J. theor. Phys. 21 (1982) 905–940.

L. von Bertalanffy, *Théorie Générale des Systèmes*. Paris: Dunod (1993).

G.M. Blackburn & M.J. Gait, *Nucleic Acids in Chemistry and Biology*, 2nd edn, pp. 210–221. Oxford: University Press (1996).

L.A. Blumenfeld, *Problems of Biological Physics*. Berlin: Springer: (1981).

I. Braslavsky et al. Sequence information can be obtained from single DNA molecules. Proc. natl Acad. Sci. USA (2003) 3960–3964.

M.G. Cacace, E.M. Landau & J.J. Ramsden, The Hofmeister series: salt and solvent effects on interfacial phenomena. Q. Rev. Biophys. 30 (1997) 241–278.

D.S. Chernavsky, Synergetics and information. Matematika kibernetika 5 (1990) 3–42 (in Russian).

R. Carnap & Y. Bar-Hillel, *An Outline of a Theory of Semantic Information*. MIT Research Laboratory of Electronics Technical Report No 247 (1952).

C. Cherry, *On Human Communication*. London: Chapman and Hall (1957).

F.H.C. Crick et al., General nature of the genetic code for proteins. Nature 192 (1961) 1227–1232.

T.G. Dewey, Algorithmic complexity of a protein, Phys. Rev. E 54 (1996) R39–R41.

T.G. Dewey, Algorithmic complexity and thermodynamics of sequence-structure relationships in proteins, Phys. Rev. E 56 (1997) 4545–4552.

R.A. Dwek & T.D. Butters (eds), Glycobiology. Chem. Rev. 102 (2002) no 2 (pp. 283 ff.).

A.W.F. Edwards, *Likelihood*. Cambridge: University Press (1972).

P. Érdi & Gy. Barna, Self-Organizing mechanism for the formation of ordered neural mappings. Biol. Cybernetics 51 (1984) 93–101.

W. Feller, *An Introduction to Probability Theory and its Applications*, 3rd edn, vol. 1. New York: Wiley (1967).

A. Fernández & H. Cendra, In vitro RNA folding: the principle of sequential minimization of entropy loss at work. Biophys. Chem. 58 (1996) 335–339.

A. Fernández & R. Scott, Dehydron: a structurally encoded signal for protein interaction. Biophys. J. 85 (2003) 1914–1928.

A. Fernández, T.R. Sosnick & A. Colubri, Dynamics of hydrogen bond desolvation in protein folding. J. mol. Biol. 321 (2002) 659–675.

A. Fernández et al., Structural defects and the diagnosis of amyloidogenic propensity. Proc. natl Acad. Sci. USA (2003) 6446–6451.

J.W. Fickett, The gene identification problem: an overview for developers. Computers Chem. 20 (1996) 103–118.

R.A. Fisher, *The Design of Experiments*, 6th edn. Edinburgh: Oliver & Boyd (1951).

S.P.A. Fodor et al., Light-directed, spatially addressable parallel chemical synthesis. Science 251 (1991) 767–773.

L.T.C. França et al., A review of DNA sequencing techniques. Q. Rev. Biophys. 35 (2002) 169–200.

H. Frauenfelder, From atoms to biomolecules. Helv. Phys. Acta 57 (1984) 165–187.

S. Galam & A. Mauger, Universal formulas for percoation thresholds. Phys. Rev. E 53 (1996) 2177–2181; *ibid.* 55 (1997) 1230–1231.

M.R. Gardner & W.R. Ashby, Connectance of large dynamic (cybernetic) systems: critical values for stability. Nature 228 (1970) 784.

M. Gell-Mann & S. Lloyd, Information measures, effective complexity, and total information. Complexity 2 (1996) 44–52.

C. Gibas & P. Jambeck, *Developing Bioinformatics Computer Skills*. Sebastopol (California): O'Reilly & Associates (2001).

I.J. Good, Statistics of language. In: *Encyclopaedia of Linguistics, Information and Control* (ed. A.R. Meetham), pp. 567–581. Oxford: Pergamon (1969).

S.J. Gould, *Ontogeny and Phylogeny*. Cambridge (Mass.): Belknap Press (1977).

P. Grassberger, Toward a quantitative theory of self-generated complexity. Int. J. theor. Phys. 25 (1986) 907–938.

S.P. Gygi, B. Rist, S.A. Gerber, F. Turecek, M.H. Gelb & R. Aebersold, Quantitative analysis of complex protein mixtures using isotope-coded affinity tags. Nature Biotechnology 17 (1999) 994–999.

R.W. Hamming, Error detecting and error correcting codes. Bell System tech. J. 26 (1950) 147–160.

R.V.L. Hartley, Transmission of information. Bell System tech. J. 7 (1928) 535–563.

B. Hartmann & R. Lavery, DNA structural forms. Q. Rev. Biophys. 29 (1996) 309–368.

H. Hillman, *The Case for New Paradigms in Cell Biology and in Neurobiology*. Lewiston: Edwin Mellen Press (1991).

R. Hooke, *Micrographia*. London: The Royal Society (1665).

B.A. Huberman & T. Hogg, Complexity and adaptation. Physica D 22 (1986) 376–384.

P. James, Protein identification in the post-genome era. Q. Rev. Biophys 30 (1997) 279–331.

H. Jeong et al., Lethality and centrality in protein networks. Nature 411 (2001) 41–42.

T.B. Jongeling, Self-organization and competition in evolution: a conceptual problem in the use of fitness landscapes. J. theor. Biol. 178 (1996) 369–373.

S. Karlin & V. Brendel, Patchiness and correlations in DNA sequences. Science 259 (1993) 677–679.

S.A. Kauffman, Emergent properties in random complex automata. Physica 10D (1984) 145–156.

E.S. Kempner & J.H. Miller, The molecular biology of *Euglena gracilis* IV. Cellular stratification by centrifuging Exp. cell Res. 51 (1968) 141–149, 150–156.

A.N. Kolmogorov, Three approaches to the quantitative definition of information. Problemy Peredachi Informatsii 1 (1965) 3–11.

A.N. Kolmogorov, Combinatorial foundations of information theory and the calculus of probabilities. Uspekhi Mat. Nauk. 38 (1983) 27–36.

A.A. Kornyshev & S. Leikin, Sequence recognition in the pairing of DNA duplexes, Phys. Rev. Lett. 86 (2001) 3666–3669.

V.G. Levich *Physicochemical Hydrodynamics*. Englewood Cliffs: Prentice-Hall (1962).

S. Kullback & R.A. Leibler, On information and sufficiency. Ann. math. Statist. 22 (1951) 79–86.

J.C. Lindon et al., Metabonomics: metabolic processes studies by NMR spectroscopy of biofluids. Concepts Magn. Reson. 12 (2000) 289–320.

P.O. Luthi et al., A cellular automaton model for neurogenesis in *Drosophila*. Physica D 118 (1998) 151–160.

S. Lloyd & H.Pagels, Complexity as thermodynamic depth. Ann. Phys. 188 (1988) 186–213.

E.H. McConkey, Molecular evolution, intracellular organization, and the quinary structure of proteins. Proc. natl Acad. Sci. USA 79 (1982) 3236–3240.

D.M. Mackay, Quantal aspects of scientific information. Phil. Mag. (ser. 7) 41 (1950) 289–311.

D.M. Mackay, Operational aspects of intellect. In: *Mechanization of thought processes*, NPL Symposium No 10, pp. 37–73. London: HMSO (1960).

B. Mandelbrot, Contribution à la théorie mathématique des jeux de communication. Publ. Inst. Statist. Univ. Paris 2 (1952) 1–124.

A.A. Markov, Statistical analysis of the text of "Eugene Onegin" illustrating the connexion with investigations into chains. Izv. Imp. Akad. Nauk. (1913) 153–162 (in Russian).

C.H. Mastrangelo, M.A. Burns & D.T. Burke, Microfabricated devices for genetic diagnostics. Proc. IEEE (August 1998) 1–15.

A.M. Mood, The distribution theory of runs. Ann. Math. Statist. 11 (1940) 367–392.

C.J. van Oss, *Forces Interfaciales en Milieux Aqueux*. Paris: Masson (1996).

R.E. Palmer, Broken ergodicity. Adv. Phys. 31 (1982) 669–735.

P.D. Patel & G. Weber, Electrophoresis in free fluid: a review of technology and agrifood applications. J. biol. Phys. Chem. 3 (2003) 60–73.

L. Peliti, Fitness landscapes and evolution. In: *Physics of Biomaterials*, eds T. Riste & D. Sherrington, pp. 287–308. Dordrecht: Kluwer (1996).

J.K. Percus, O.E. Percus & A.S. Perelson, Predicting the size of the T-cell receptor and antibody combining region from consideration of efficient self-nonself discrimination. Proc. natl Acad. Sci. USA 90 (1993) 1691–1695.

M. Planck, The concept of causality. Proc. Phys. Soc. 44 (1932) 529–539.

G.H. Pollack, *Cells, Gels and the Engines of Life*. Seattle: Ebner (2001).

O. Popescu & G.N. Misevic, Self-recognition by proteoglycans. Nature 386 (1997) 231–232.

D.J. Raine & V.J. Norris, Network structure of metabolic pathways. J. biol. Phys. Chem. 1 (2002) 89–94.

J.J. Ramsden, The photolysis of small silver halide particles. Proc. R. Soc. Lond. A 392 (1984) 427–444.

J.J. Ramsden, Computing photographic response curves. Proc. R. Soc. Lond. A406 (1986) 27–37.

J.J. Ramsden, Experimental methods for investigating protein adsorption kinetics at surfaces. Q. Rev. Biophys. 27 (1994) 41–105.

J.J. Ramsden, Kinetics of protein adsorption. In: *Biopolymers at Interfaces* (ed. M. Malmsten), pp. 321–361. New York: Dekker (1998).

J.J. Ramsden & M. Grätzel, Formation and decay of methyl viologen radical cation dimers on the surface of colloidal CdS. Chem. phys. Lett. 132 (1986) 269–272.

J.J. Ramsden & J. Dreier, Kinetics of the interaction between DNA and the type IC restriction enzyme *Eco*R124/3I. Biochemistry 35 (1996) 3746–3753.

J.J. Ramsden & J. Vohradský, Zipf-like behavior in procaryotic protein expression. Phys. Rev. E 58 (1998) 7777–7780.

J.J. Ramsden, D.J. Roush, D.S. Gill, R.G. Kurrat & R.C. Willson, Protein adsorption kinetics drastically altered by repositioning a single charge. J. Amer. chem. Soc. 117 (1995) 8511–8516.

A. Rényi, Kémiai reakciók tárgyalása a sztochasztikus folyamatok elmélete segítségével. Magy. tud. Akad. mat. Kut. Int. Közl. 2 (1953) 83–101.

A. Rényi, *Probability Theory*. Budapest: Akadémiai Kiadó (1970).

J. Rissanen, *Stochastic Complexity in Statistical Enquiry*. Singapore: World Scientific (1989).

J.C. Robinson, All possible chaotic dynamics can be approximated in three dimensions. Nonlinearity 11 (1998) 529–545.

V. Ye. Ruzhentsev, The problem of transition in palaeontology. Int. Geology Rev. 6 (1964) 2204–2213.

F. Sanger, Determination of nucleotide sequences in DNA. Bioscience Reports 1 (1981) 3–18.

M.A. Savageau, Comparison of classical and autogenous systems of regulation in inducible operons. Nature 252 (1974) 546–549.

F. Scheller & F. Schubert, *Biosensoren*. Berlin: Akademie-Verlag (1989).

P. Schuck, Kinetics of ligand binding to receptors immobilized in a polymer matrix, as detected with an evanescent wave biosensor. I. A computer simulation of the influence of mass transport. Biophys. J. 70 (1996) 1230–1249.

C.E. Shannon, A mathematical theory of communication. Bell System tech. J. 27 (1948) 379–423.

C.E. Shannon & W. Weaver, *The Mathematical Theory of Communication*. Urbana: University of Illinois Press (1949).

R. Shaw, Strange attractors, chaotic behaviour, and information flow. Z. Naturforsch. 36a (1981) 80–112.

A.K. Solomon, Red cell membrane structure and ion transport. J. gen. Physiol. 43 (1960) 1–15.

G. Sommerhoff, *Analytical Biology*, London: Oxford University Press (1950).

G. Stent, Explicit and implicit semantic content of the genetic information. In: *The Centrality of Science and Absolute Values*, 4th Int. Conf. on the Unity of the Sciences, vol. 1, pp. 261–277. New York: International Cultural Foundation (1975).

M.C.R. Symons, Water structure and reactivity. Acc. chem. Res. 14 (1981) 179–187.

W.H. Thorpe, Purpose and mechanism in biology. In: *The Centrality of Science and Absolute Values*, 4th Int. Conf. on the Unity of the Sciences, vol. 1, pp. 623–634. New York: International Cultural Foundation (1975).

W.H. Thorpe, *Purpose in a World of Chance*. Oxford: University Press (1978).

N. Tkemaladze, On the problems of an automated system of pattern recognition with learning. J. Biol. Phys. Chem. 2 (2002) 80–84.

G.M. Tomkins, The metabolic code. Science 189 (1975) 760–763.

R. Tureck, Cells, functions, relationships in musical structure and performance. Proc. R. Inst. 67 (1995) 277–318.

R.A. VanBogelen et al., Mapping regulatory networks in microbial cells. Trends Microbiol. 7 (1999) 320–327.

S.B. Volchan, What is a random sequence? Amer. math. Monthly 109 (2002) 46–63.

J. Vohradský, Neural network model of gene expression. FASEB J. 15 (2001) 846–854.

J. Vohradský & J.J. Ramsden, Genome resource utilization during procaryotic development. FASEB J. 15 (2001) 2054–2056.

R.F. Voss, Evolution of long-range fractal correlations and $1/f$ noise in DNA base sequences. Phys. Rev. Lett. 68 (1992) 3805–3808.

B.L. van der Waerden, Beweis einer Baudet'schen Vermutung. Nieuw. Arch. Wiskunde 15 (1927) 212–216.

A. Wagner & D.A. Fell, The small world inside large metabolic networks. Proc. R. Soc. Lond. B 268 (2001) 1803–1810.

J.C. Wang, Moving one DNA double helix through another by a type II DNA topoisomerase. Q. Rev. Biophys. 31 (1998) 107–144.

D.J. Watts & S.H. Strogatz, Collective dynamics of 'small-world' networks. Nature 393 (1998) 440–442.

S.H. White, Global statistics of protein sequences. A. Rev. Biophys. biomol. Structure 23 (1994) 407–439.

S. Wolfram, Statistical mechanics of cellular automata. Rev. mod. Phys. 55 (1983) 601–644.

S. Wright, Character change, speciation and the higher taxa. Evolution 36 (1982) 427–443.

Index